数据库 MySQL
入门到精通"私房菜"

王丁丁 编著

北京航空航天大学出版社

内 容 提 要

目前，众多大型互联网、电信、制造业等用户已经普遍使用 MySQL 构建其业务，国内众多能源、广电、高科技、软件开发企业也渐渐将 MySQL 作为数据库平台的优先选择。

本书从专业技术视角入手讲解 MySQL 系统性的相关知识，内容涵盖 MySQL8.0 版本内容的安装部署、基础运维、体系架构以及高可用等。学习本书可帮助相关人员快速掌握 MySQL 数据库技术知识，提高处理实际问题的能力。

本书注重实用性、可操作性，特别适合数据库优化、DBA 开发、DBA 运维、IT 管理等人员使用，另外，缺乏现场实施经验、想要体系化掌握 MySQL、考取 MySQL OCP 的读者，也可参考阅读。

图书在版编目（CIP）数据

数据库 MySQL 入门到精通"私房菜" / 王丁丁编著．
北京：北京航空航天大学出版社，2024.7. --ISBN 978-7-5124-4480-5

Ⅰ．TP311.132.3

中国国家版本馆 CIP 数据核字第 2024QK3826 号

版权所有，侵权必究。

数据库 MySQL 入门到精通"私房菜"
王丁丁　编著
策划编辑　杨晓方　　责任编辑　杨晓方

*

北京航空航天大学出版社出版发行
北京市海淀区学院路 37 号（邮编 100191）　http://www.buaapress.com.cn
发行部电话：(010)82317024　传真：(010)82328026
读者信箱：copyrights@buaacm.com.cn　邮购电话：(010)82316936
北京一鑫印务有限责任公司印装　各地书店经销
开本：710×1 000　1/16　印张：19.75　字数：410 千字
2024 年 8 月第 1 版　2024 年 8 月第 1 次印刷
ISBN 978-7-5124-4480-5　定价：99.00 元

若本书有倒页、脱页、缺页等印装质量问题，请与本社发行部联系调换　联系电话：(010)82317024

前　言

MySQL 是一种可靠、可扩展且易于使用的开源关系数据库系统，其功能包括 SQL 标准命令以及事务和 ACID 合规性（代表原子性、一致性、隔离性和持久性）。MySQL 一直引领着开源数据库的发展。

MySQL 的具体优势如下：

（1）运行速度快：MySQL 中使用了极快的 B 树磁盘表（MyISAM）和索引压缩，通过使用优化的单扫描多连接，能够极快地实现连接。SQL 的函数使用高度优化的类库实现，运行速度极快。MYSQL 在 innodb 引擎出来后，事务处理方面可满足大部分场景。

（2）高灵活性：MySQL 既可以嵌入式应用于程序中，也可支持数据仓库、内容索引和部署软件、高可用的冗余系统、在线事务处理系统（OLTP）等各类应用类型。

（3）成本低：MySQL 是开源的，无版权制约，使用方便，这使得 MySQL 可以更加灵活地满足各种需求，降低使用成本。

（4）优异的性能：MySQL 的存储引擎架构将查询处理和其他系统任务、数据的存储提取相分离。这种处理和存储分离的设计可以在使用时根据性能以及其他需求来选择数据存储的方式，使得 MySQL 在处理大量数据时更加高效。

DBA 属于运维范畴，涉猎范围广泛：DB（Oracle、MySQL、PG）、Linux、硬件、网络、脚本（Python、Shell）、监控（Zabbix、Prometheus）等，谁掌握了数据，确保了数据的安全，就会在快速发展的信息化中独占鳌头，DBA 工作会让你永远保持清醒的思维。作者积累的 10 余年 MySQL 数据库运维工作经验都在本书中有详细的介绍和说明。

由于作者的水平有限，书中难免会出现一些错误或者不准确的地方，恳请广大读者批评指正，拓展资料获取方式通过微信号 jem_db 联系。

在阅读本书的过程中，如果读者有任何宝贵意见，均可以通过微信公众号"IT 邦德"进行后台留言或发邮件至 2243967774@qq.com 与作者沟通和交流。

本书约定

本书实战内容大多源自企业日常运维，客户实施案例，不做特殊说明，均为 MySQL8 版本。

本书使用不同的底色区分代码和一般正文内容，对于重要的概念也采用了加粗处理。

作者从实际工作问题出发，由易到难进行讲解，为读者构建完整的 MySQL 架构体系。

在本书的最后，作者对所有的内容做了归纳总结，形成了一个完善体系化的MySQL技能树帮助读者掌握技能。

致谢

感谢我的爱人，是她在背后默默地支持我顺利完成了本书的创作。

感谢我的 ACE 的朋友们，他们以专业的视角帮忙审稿，提出了许多宝贵的意见。

感谢北京航空航天大学出版社的大力支持和帮助，使我的作品得以出版。

感谢广大读者朋友的厚爱和支持。

目 录

Part 1 基础篇 ... 1

第 1 章 数据库基本概念 ... 2
- 1.1 数据库三大范式 ... 2
- 1.2 事务的 ACID ... 4
- 1.3 存储引擎 ... 6
- 1.4 MySQL 的优势 ... 7

第 2 章 MySQL 服务概述 ... 9
- 2.1 MySQL 发行版 ... 9
- 2.2 MySQL 发展史 ... 11
- 2.3 MySQL 安装简介 ... 12

第 3 章 MySQL 安装及卸载 ... 15
- 3.1 Windows 环境 MSI 图形安装 ... 15
- 3.2 Windows 环境二进制安装 ... 21
- 3.3 Linux PRM 包安装 MySQL ... 24
- 3.4 Linux 二进制多版本部署 MySQL ... 30
- 3.5 Linux 源码安装 MySQL ... 40
- 3.6 MySQL 版本升级 ... 45
- 3.7 MySQL 客户端工具 ... 50

Part 2 运维篇 ... 55

第 4 章 DBMS 基本管理 ... 56
- 4.1 数据类型 ... 56
- 4.2 数据库管理 ... 59
- 4.3 表的管理 ... 60
- 4.4 用户管理 ... 67
- 4.5 SQL 实战 ... 77
- 4.6 函数的用法 ... 91
- 4.7 约 束 ... 96
- 4.8 存储过程 ... 107
- 4.9 触发器 ... 114

- 4.10 事件(event) …… 116
- 4.11 Prometheus 监控 …… 119
- 4.12 Zabbix 监控 …… 124

Part 3 体系架构篇 …… 135

第 5 章 物理结构 …… 138
- 5.1 参数文件 …… 138
- 5.2 日志文件 …… 141
- 5.3 MySQL 表结构文件 …… 154
- 5.4 其他文件 …… 155

第 6 章 存储引擎结构 …… 157
- 6.1 内存结构 …… 157
- 6.2 磁盘结构 …… 162
- 6.3 MySQL 逻辑存储 …… 163
- 6.4 表空间运维 …… 165
- 6.5 分区表 …… 166
- 6.6 线程结构 …… 173

Part 4 备份恢复篇 …… 177

第 7 章 逻辑备份 …… 178
- 7.1 MySQL 导入导出 …… 178
- 7.2 Mysqldump 逻辑备份恢复 …… 182
- 7.3 Mysqlpump …… 185

第 8 章 物理备份 …… 187
- 8.1 PXB 介绍及部署 …… 187
- 8.2 PXB 备份恢复 …… 188

第 9 章 MySQL 误操作恢复 …… 192
- 9.1 Mysqlbinlog 恢复误删除 …… 192
- 9.2 第三方工具 …… 195

Part 5 高可用 …… 197

第 10 章 主从复制 …… 198
- 10.1 主从复制简介 …… 198
- 10.2 主从复制优点 …… 198
- 10.3 主从复制原理 …… 199
- 10.4 主从复制方式 …… 200

10.5	传统异步主从复制部署	203
10.6	主从复制维护	206
10.7	单主2从 GTID 复制	208

第 11 章 MySQL Router 读写分离及负载均衡 213

11.1	MySQL Router 简介	213
11.2	MySQL Router 工作流程	213
11.3	读写分离＋负载均衡	214

第 12 章 高可用 MHA 架构 220

12.1	MHA 简介	220
12.2	架构规划	220
12.3	安装 MySQL8	221
12.4	一主2从 GTID 同步	222
12.5	互信设置	225
12.6	安装 MHA 软件	227
12.7	配置 MHA	227
12.8	创建脚本	229
12.9	MHA 服务启动	232
12.10	MHA 故障转移	233

第 13 章 高可用之多源复制 236

13.1	多源复制简介	236
13.2	多源复制部署	236

第 14 章 高可用之 MGR 架构 244

14.1	MGR 简介	244
14.2	MGR 特点	244
14.3	MGR 部署	245
14.4	MGR 单主模式	249
14.5	多主和单主模式切换	251

第 15 章 双主＋Keepalived 单点故障切换 254

15.1	Keepalived 简介	254
15.2	双主复制要点	255
15.3	双主部署	255

第 16 章 高可用之 PXC 架构 263

16.1	PXC 介绍	263
16.2	PXC 的优缺点	264
16.3	PXC 的原理	264
16.4	PXC 的重要概念	265

16.5　PXC 高可用集群部署 …………………………………… 268
　　16.6　HAProxy ………………………………………………… 277
第 17 章　高可用之分库分表 ……………………………………… 282
　　17.1　分库分表简介 …………………………………………… 282
　　17.2　分库分表的目的 ………………………………………… 282
　　17.3　常用数据库中间件 ……………………………………… 283
　　17.4　分库分表的类型 ………………………………………… 283
　　17.5　小　结 …………………………………………………… 284
附录 1　8.0.30 or Higher 新特性 ………………………………… 286
　　1. Redo Log ……………………………………………………… 286
　　2. GIPK ………………………………………………………… 287
　　3. 多级别的 ORDER BY or LIMIT …………………………… 290
　　4. innodb_doublewrite ………………………………………… 291
　　5. Mysqldump …………………………………………………… 291
附录 2　性能优化 …………………………………………………… 292
　　1. 统计信息 ……………………………………………………… 292
　　2. 统计信息不准确的处理 ……………………………………… 292
　　3. 执行计划查看 ………………………………………………… 293
附录 3　MySQL 技能树 …………………………………………… 295

Part 1　基础篇

第1章 数据库基本概念

本章介绍了数据库的相关基本概念,这些概念是学习 MySQL 的先决条件。

1.1 数据库三大范式

为了建立冗余较小、结构合理的数据库,设计数据库时必须遵循一定的规则,在关系型数据库中这种规则就被称为范式。范式是符合某一种设计要求的总结。因此要设计一个结构合理的关系型数据库,就必须要满足下面这 3 大范式。

1.1.1 第一范式(不可再分)

数据库表中的字段都是单一属性的,不可再分。这个单一属性由基本类型构成,包括整型、实数、字符型、逻辑型、日期型等。第一范式(1NF)主要是确保数据表中每个字段的值必须具有原子性,也就是说数据表中每个字段的值为不可再次拆分的最小数据单元。

【案例】确保每一列的原子性

表 1-1 中所列的表实际上就不满足 1NF,因为班级这列是可以继续被拆分的。

表 1-1 不满足 1NF 的班级信息表

姓名	班级
李四	计算机 2208

表 1-2 所列的表中不能有可以被继续拆分的列,即表中的每一个属性列都具有原子性。

表 1-2 满足 1NF 的班级信息表

姓名	专业	班号
李四	计算机	2208

1.1.2 第二范式(消除部分依赖)

第二范式(2NF)是指在第一范式的基础上,确保数据表中除了主键之外的每个字段都必须依赖主键,非主键列完全依赖于主键,而不能依赖于主键的一部分。

【案例】第二范式的数据表的设计

不是所有属性字段都完全依赖联合主键的,它们或许只依赖主键中的一部分,这种部分依赖的关系是不满足 2NF 的。表 1-3 中的例子中,由于商品的名称和价格字段不依赖于商品类别的主键 id,所以不符合第二范式,因此需要拆表。可以将其修改成表 1-4 category 和表 1-5 goods 的设计形式,商品信息 goods 表通过商品类别 id 字段与数据表 category 中商品类别 category_id 字段进行关联。

表 1-3 不满足 2NF 的表

字段名称	字段类型	是否主键	说明	示例
id	INT	是	商品类别主键	1001
category	VARCHAR(50)	否	商品类别	日用品
goods	VARCHAR(50)	否	商品名称	牙刷
price	DECIMAL(2,1)	否	商品价格	26.6

符合第二范式的 category 数据表的设计,如表 1-4 所列。

表 1-4 category 数据表

字段名称	字段类型	是否主键	说明	示例
id	INT	是	商品类别主键	1001
category	VARCHAR(30)	否	商品类别	日用品

符合第二范式的 goods 数据表的设计,如表 1-5 所列。

表 1-5 goods 数据表

字段名称	字段类型	是否主键	说明	示例
id	INT	是	商品主键	100001001
category_id	VARCHAR(30)	否	商品类别 id	1001
goods	VARCHAR(30)	否	商品名称	牙刷
price	DECIMAL(2,1)	否	商品价格	23.6

1.1.3 第三范式(消除间接依赖)

在第二范式的基础上,非主键列只依赖于主键,不依赖于其他非主键。第三范式(3NF)需要确保数据表中的每一列数据都和主键直接相关,而不能间接相关。

【案例】第三范式的数据表的设计

表 1-6 中主键是"学号",直接依赖于"学号"的有"姓名"和"课程号"。"课程名称"直接依赖于"课程号",间接依赖于"学号"。因此,我们需要为课程号和课程名称单独创建一个表出来。

表 1-6 不满足 3NF 的表

学号	姓名	课程号	课程名称
22089	李四	360	计算机

符合第三范式的学生表的设计,如表 1-7 所列。

表 1-7 满足 3NF 的学生表

学号	姓名	课程号
22089	李四	360

符合第三范式的课程表的设计,如表 1-8 所列。

表 1-8 满足 3NF 的课程表

课程号	课程名称
360	计算机

小　结

三大范式只是一般设计数据库的基本理念,其可以建立冗余较小、结构合理的数据库。当然如果有特殊情况,要特殊对待,数据库设计最重要的是看业务需求和性能,顺序是:需求＞性能＞表结构,所以不能一味追求范式建立数据库关联。要尽量遵守三范式,如果不遵守,必须有足够的理由,事实上,我们经常会为了性能而选择不太友好的数据库设计。

1.2　事务的 ACID

关系型数据库需要遵循 ACID 规则。ACID 是指数据库管理系统(DBMS)在写入或更新资料的过程中,为保证事务(transaction)是正确可靠的而必须具备的 4 个特性:原子性(atomicity 或称不可分割性)、一致性(consistency)、隔离性(isolation 又称独立性)、持久性(durability)。

1.2.1　原子性(atomicity)

一个事务中的所有操作,要么全部完成,要么全部不完成,不会结束在中间某个环节。如果事务在执行过程中发生错误,会被恢复(rollback)到事务开始前的状态,就像这个事务从来没有执行过一样,事务的所有操作要么全部成功,要么全部回滚。

如图 1-1 所示,总共 1 000 元钱,A 和 B 发生转账交易,这个过程包含两个步骤。

A:850 - 150 = 700,A 原来有 850 元,最终余额为 700 元。
B:150 + 150 = 300,B 原来有 150 元,最终余额为 300 元。
原子性表示,这两个步骤一起成功,或者一起失败,不能只发生其中一个动作,如图 1-1 所示。

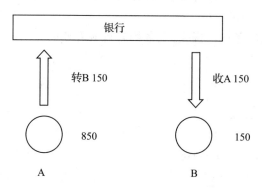

图 1-1 满足原子性的银行交易

1.2.2 一致性(consistency)

在事务开始之前和事务结束以后,数据库的完整性没有被破坏。这表示写入的数据必须完全符合所有的预设规则,这包含数据的精确度、串联性以及后续数据库可以自发性地完成预定的工作,即事务前后数据的完整性必须保持一致。图 1-1 所示发生的转账交易,一致性表示事务完成后,如果事务提交成功,则 A 账户减金额,B 账户加对应的金额,数据库总体金额不变,只是载体变了。如果事务出错,则整体回滚,无论到了上面的哪个步骤,A 和 B 的数据都会回到事务开启前的状态,保证数据的始终一致,符合逻辑运算。

1.2.3 隔离性(isolation)

隔离性是指事务内部的操作与其他事务是隔离的,并发执行的各个事务之间不能互相干扰。隔离性追求的是并发情形下事务之间互不干扰。为简单起见,我们主要考虑使用较简单的读操作和写操作,那么隔离性的探讨主要可以分为两个方面:
(1)一个事务写操作对另一个事务写操作的影响:锁机制保证隔离性。
(2)一个事务写操作对另一个事务读操作的影响:MVCC 保证隔离性。
如图 1-2 所示,针对多个用户同时操作时,主要排除其他事务对本次事务的影响。

1.2.4 持久性(durability)

一个事务被提交之后,它对数据库中数据的改变是持久的,即使数据库发生故障也不应该对其有任何影响,事务结束后的数据不会随外界原因而丢失。

以图1-2所示的银行交易为例。

操作前 A:850,B:150。

操作后 A:700,B:300。

如果在操作前(事务还没有提交)服务器宕机或者断电,那么重启数据库以后,数据状态应该为 A:850,B:150。

如果在操作后(事务已经提交)服务器宕机或者断电,那么重启数据库以后,数据状态应该为 A:700,B:300。

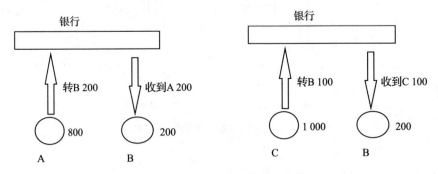

图1-2 满足隔离性的转账交易

小　结

事务是并发控制单位,也是用户定义的一个操作序列;这些操作要么都做,要么都不做,它们是一个不可分割的工作单位。事务通常以 BEGIN TRANSACTION 开始,以 COMMIT 或 ROLLBACK 结束。事务中的所有操作要么全部执行,要么都不执行。如果事务没有原子性的保证,那么在发生系统故障的情况下,数据库就有可能处于不一致状态。因而,事务的原子性与一致性是密切相关的。

1.3　存储引擎

存储引擎是存储系统的发动机,直接决定了存储系统的性能和功能。存储引擎负责管理数据具体在介质上存储和检索。不同的存储引擎的优劣势不同,我们需要了解不同存储引擎的优缺点,才能根据需求选择合适的存储引擎来管理数据,从而来提高存储系统的性能,以下介绍几种常用的存储引擎。

1.3.1　InnoDB 存储引擎

InnoDB 是一种兼顾高可靠性和高性能的通用存储引擎,在 MySQL5.5 之后,InnoDB 是默认的 MySQL 存储引擎,支持事务处理、外键以及崩溃修复能力和并发控制。如果需要对事务的完整性要求比较高(比如证券),要求实现并发控制(比如售票),那选择使用 InnoDB 有很大的优势。如果需要频繁更新、删除操作的数据库,也可以选择 InnoDB,因为它支持事务的提交(commit)和回滚(rollback)。

1.3.2 MyISAM 存储引擎

MyISAM 是 MySQL 早期的默认存储引擎不支持事务,也不支持外键,但支持表锁,不支持行锁,占用空间小,访问速度快。插入数据快,空间和内存使用比较低。如果使用表主要是用于插入新记录和读出记录,那么选择 MyISAM 能实现处理高效率,一般用于以读/写插入为主的应用程序,比如博客系统、新闻门户网站。

1.3.3 Memory 存储引擎

Memory 引擎的表数据是存储在内存中的,由于受硬件问题或断电问题的影响,只能将这些表作为临时表或缓存使用。如果需要很快读/写速度,且对数据的安全性要求较低,那么可以选择 MEMOEY,它对表的大小有要求,如不能建立太大的表。

1.3.4 ARCHIVE 存储引擎

ARCHIVE 存储引擎非常适合存储大量独立的,作为历史记录的数据。ARCHIVE 提供了压缩功能,拥有高效的插入速度,但是这种引擎不支持索引,所以查询性能较差。

小　结

在 MySQL 数据库中可以使用多种存储引擎的表,如果一个表用于比较高的事务处理,那么可以选择 InnoDB。这个数据库中可以将查询要求比较高的表选择 MyISAM 存储。如果该数据库需要一个用于查询的临时表,那么可以选择 MEMORY 存储引擎。

1.4　MySQL 的优势

目前,MySQL 是世界上较流行的开源数据库。无论您是快速增长的 web 资产、技术 ISV 还是大型企业,MySQL 都可以经济高效地帮助您交付高性能、可扩展的数据库应用程序。

MySQL 是一个关系型数据库管理系统,由瑞典 MySQL AB 公司开发,目前属于 Oracle 旗下产品。关系数据库将数据保存在不同的表中,而不是将所有数据放在一个大仓库内,这样就提升了速度,并提高了灵活性。

当下,MySQL 在全球普及,用户遍布互联网、电信、能源、交通、高科技、制造业、科研和军事领域，政府机构、人口、社保、教育、医疗系统，从武器研发到警察、武装部队，很多都采用 MySQL 作为后台数据库。一方面 MySQL 的开源属性可为使用者节约大量的软件采购费用，另一方面 MySQL 的代码公开、透明，也使得安全风险较低，其特点如下：

（1）MySQL 社区版是开源的，所以使用者不需要支付额外的费用，其官网主页如图 1-3 所示。

（2）MySQL 支持大型的数据库，可以处理上千万条记录的大型数据库。

（3）MySQL 使用标准的 SQL 数据语言形式。

（4）MySQL 允许用于多个系统，并且支持多种语言，这些编程语言包括 C、C++、Python、Java、Perl、PHP、Eiffel、Ruby、Rust。

（5）MySQL 是可以定制的，采用了 GPL 协议，使用者可以修改源码来开发自己的 MySQL 系统。

从技术角度讲，我认为 MySQL 最大的优势的不在于经常被媒体称道的"快速"和"轻量"，而是前后端分离，对于资源非常有限的早期 MySQL，这个选择是需要勇气的，但是这奠定了未来几十年间 MySQL 的技术优势。面对层出不穷的各种挑战，独立的存储引擎是其最大的竞争优势，并且，这个壁垒是竞争对手难以跨越的，它根植于这个软件产品的根本设计思想。MySQL 这个开源软件适用于各个技术水平的开发者。使用者完全可以从熟悉 MySQL 开始，根据自己的需要，选择适合自己的方向深入学习。

图 1-3　MySQL 官网主页

第 2 章　MySQL 服务概述

本章为 MySQL 的服务概述,对其发展史及发行版本、安装部署类型做了详细介绍。

2.1　MySQL 发行版

2.1.1　MySQL 官方版

MySQL 官方版分为社区版和企业版,企业版可定制,性能较强,提供了很多的功能和工具,有完善的技术支持服务。社区版完全免费,但不提供技术支持。

1. 企业版

MySQL 企业版(商业版)是由 MySQL 公司内部发布,同时参考社区版的先进代码功能和算法,是 MySQL 公司的盈利产品,需要付费才能使用及享用服务支持,其稳定性和可靠性无疑都是最好的,但不遵守 GPL 协议。

企业版经过严格测试认证,稳定、安全、可靠,性能也比社区版好。企业版使用商业的编译器对代码进行编译和优化,源代码有规律且稳定性和执行效率高,各版本平台绑定优化,同时包含企业级图形监控软件、服务和支持,可以监控软件运行状态,技术预警,出现问题后可根据源码编排规律和资深 MySQL 数据库专家及时查找和修正,使技术风险降到最低,定期的升级支持包可以解决软硬件兼容性问题。

2. 社区版

MySQL 社区版则由分散在世界各地的 MySQL 开发者、爱好者以及用户参与开发与测试,并完成软件代码的管理,测试工作。社区也会设立 BUG 汇报机制,收集用户遇到 BUG 问题,相比商业版,社区版的开发及测试环境没有那么严格,遵守 GPL 协议。

社区版在技术方面会加入许多新的未经严格测试的特性,以便从广大社区用户那里得到反馈和修正。社区版源码无规律,很多社区用户都可以补充和修正,社区版未经各个专有系统平台的压力测试和性能测试,所以社区版在当今高速发展的软件和硬件体系的兼容性方面都可能存在技术风险。

2.1.2　Percona MySQL

Percona Server 是 MySQL 重要的分支,它基于 InnoDB 存储引擎的基础,提升了性能和易管理性,最后形成了增强版的 XtraDB 引擎,可以更好地发挥服务器硬件的性能,从目前公布的数据来看,Percona MySQL 性能要优于 MySQL 官方社区版的。

Percona Server 由领先的 MySQL 咨询公司 Percona 发布。

Percona Server 是一款独立的数据库产品,其可以完全与 MySQL 兼容,可以在不更改代码的情况了下将存储引擎更换成 XtraDB。

Percona 团队的最终声明是:Percona Server 是由 Oracle 发布的最接近官方 MySQL Enterprise 发行版的版本。因此与其他更改了大量基本核心 MySQL 代码的分支有所区别。

Percona Server 的一个缺点是他们自己管理代码,不接受外部开发人员的意见,以这种方式确保他们对产品中所包含功能的控制。

Percona 提供了高性能 XtraDB 引擎,还提供 PXC 高可用解决方案,并且附带了 perconatoolkit 等 DBA 管理工具箱。

2.1.3　MariaDB

MariaDB 是由 MySQL 初始创建者在 MySQL 被 Oracle 收购之后,又独立成立了一家公司,来开发的数据库。其是由 MySQL5.5 源代码为基础发展起来的。MariaDB 并不能完全和 MySQL 官方版完全兼容,但大多数功能是兼容的。我们可以很容易将数据从 MySQL 迁移到 MariaDB 上。MariaDB 主要由开源社区维护,采用 GPL 授权许可,其目的是完全兼容 MySQL,包括 API 和命令行,所以它轻松成为 MySQL 的代替品,官方主页如图 2-1 所示。

图 2-1　MariaDB 官方主页

知识拓展:Oracle 收购 MySQL 之后,MySQL 创始人之一 Monty 担心 MySQL 数据库发展的未来(开发缓慢,封闭,可能会被闭源),就创建了一个分支 MariaDB,默认使用全新的 Maria 存储引擎,它是原 MyISAM 存储引擎的升级版本。MariaDB 由 MySQL 的创始人 Michael Widenius 主导开发,他早前曾以 10 亿美元的价格,将自己创建的公司 MySQL AB 卖给了 SUN,此后,随着 SUN 被甲骨文收购,MySQL 的所有权也落入 Oracle 的手中，MariaDB 名称来自 Michael Widenius 的女儿 Maria 的名字。

2.1.4 各个发行版之间的区别

MySQL 各个发行版之间的基础对比见表 2-1 所列。

表 2-1 MySQL 各个发行版之间的基础对比

类别	MySQL	Percona MySQL	MariaDB
是否开源	社区版开源	开源	开源
事务型存储引擎	InnoDB	XtraDB	XtraDB
监控工具	企业版监控工具,社区版不提供	Percona Monitor 工具	Monyog
防火墙	企业版提供	ProxySQL FireWall	MaxScale FireWall

(1) InnoDB 属于 Oracle 公司,不对外提供,因此对于事务型存储引擎,Percona MySQL 和 MariaDB 使用的都是由 percona 公司开发的 XtraDB,XtraDB 和 InnoDB 完全兼容。

(2) 对于监控工具来说,尽管 MySQL 社区版不提供监控工具,但使用 Percona Monitor 工具或 Monyog 都可以实现对 MySQL 社区版的数据监控。

(3) 防火墙方面,MySQL 社区版是没有这个概念的,MySQL 只在企业版提供防火墙功能。Percona MySQL 和 MariaDB 都是通过其数据库中间层产品来间接提供防火墙功能。但值得一提的是,MariaDB 的 MaxScale FireWall 并不是开源的。

2.2 MySQL 发展史

下面简单回顾下相关的一些版本信息

1996 年 MySQL1.0 发布。它的历史可以追溯到 1979 年,作者 Monty 用 BASIC 设计的一个报表工具。

1996 年 10 月 3.11.1 发布,MySQL 没有 2.x 版本。

2000 年 ISAM 升级成 MyISAM 引擎,MySQL 开源。

2003 年 MySQL 4.0 发布,集成 InnoDB 存储引擎。

2005 年 MySQL 5.0 版本发布,提供了视图、存储过程等功能。

2008 年 MySQL AB 公司被 Sun 公司收购,进入 Sun MySQL 时代。

2009 年 Oracle 收购 Sun 公司,进入 Oracle MySQL 时代。

2010 年 MySQL 5.5 发布,InnoDB 成为默认的存储引擎。

2016 年 MySQL 发布 8.0.0 版本。

2024 年 7 月 MySQL 发布 9.0.0 创新版本。

2.3 MySQL 安装简介

2.3.1 MySQL 支持的操作系统

MySQL 不但支持 Ubuntu、RedHat、CentOS、Windows、macOS 等操作系统，同时还支持虚拟化环境中部署。

表 2-2 仅展示部分，MySQL5.7～8.0 各版本支持情况请访问 MySQL 官网进行查询。

表 2-2　MySQL Supported Platforms

Operating System	Architecture	8.4 LTS	8.0	5.7
Oracle Linux / Red Hat /CentOS / Rocky Linux				
Oracle Linux 9 / Red Hat Enterprise Linux 9 / Rocky Linux 9	x86_64，ARM 64	•	•	
Oracle Linux 8 / Red Hat Enterprise Linux 8 / CentOS 8 / Rocky Linux 8	x86_64，ARM 64	•	•	
Oracle Linux 7 / Red Hat Enterprise Linux 7 / CentOS 7	x86_64，ARM 64	•	•	
Oracle Linux 6 / Red Hat Enterprise Linux 6 / CentOS 6	x86_32，x86_64		•	•
Oracle Solaris				
Solaris 11（Update 4＋）	SPARC_64		•	•
SUSE				
SUSE Enterprise Linux 15 /OpenSUSE 15（15.5）	x86_64	•	•	
SUSE Enterprise Linux 12（12.5＋）	x86_64		•	
Debian				
Debian GNU/Linux 12	x86_64	•	•	
Microsoft Windows Server				
Microsoft Windows 2022 Server	x86_64	•	•	
Microsoft Windows 2019 Server	x86_64	•	•	
Microsoft Windows 2016 Server	x86_64	•		•
Microsoft Windows 2012 Server R2	x86_64	•		•

续表 2-2

Operating System	Architecture	8.4 LTS	8.0	5.7
Microsoft Windows				
Microsoft Windows 11	x86_64	•	•	
Microsoft Windows 10	x86_64		•	•
Apple				
macOS 14	x86_64，ARM 64	•	•	
macOS 13	x86_64，ARM 64	•	•	
Canonical				
Ubuntu 24.04 LTS	x86_64	•	•	
Ubuntu 22.04 LTS	x86_64	•	•	
Ubuntu 20.04 LTS	x86_64	•	•	
Ubuntu 18.04 LTS	x86_32，x86_64	•	•	•

2.3.2 MySQL 安装方式

MySQL 安装方式如图 2-2 所示。

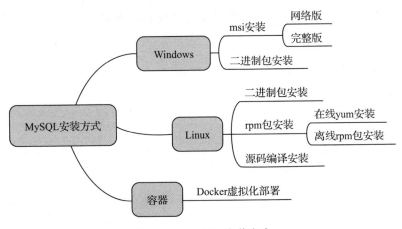

图 2-2 MySQL 安装方式

(1)在 Windows 下可以使用二进制包和 MSI 图形化包来安装。
(2)在 Linux 下可以使用 RPM、二进制、源码 3 种方式来安装,如图 2-3 所示。
(3)新手在刚入门时难免遇到安装各种各样的坑,Docker 部署可以避过对环境的依赖。

图 2-3 Linux 下 MySQL 安装的 3 种方案的比较

类别	RPM (Redhat Package Manage)	二进制 (Binary Package)	源码 (Source Package)
优点	安装简单,适合初学者学习使用 生产环境一般用这种方式	安装简单;可以安装到任何路径下,灵活性好;一台服务器可以安装多个 MySQL	在实际安装的操作系统中可根据需要定制编译,最灵活;性能最好;一台服务器可以安装多个 MySQL
缺点	需要单独下载客户端和服务器;安装路径不灵活,默认路径不能修改,一台服务器只能安装一个 MySQL	已经经过编译,性能不如源码编译的好;不能灵活定制编译参数	安装过程较复杂;编译时间长
下载选项	Red Hat Enterprise Linux / Oracle Linux	Linux - Generic	Source Code

小 结

从关系型数据库的功能来讲,MySQL 不是最强大的产品,但是 MySQL 在很多年发展历程中,每个版本成熟都离不开大量真实业务场景的打磨。企业中大多数为 Linux 服务器,所以读者要根据实际的环境选择部署的方式。

第 3 章 MySQL 安装及卸载

本章分别介绍在不同的操作系统下部署 MySQL 服务及数据库升级，可满足多场景下的工作需求，且将遇到的问题点做了详细的阐述，对于部署实施的 DBA 非常实用。

3.1 Windows 环境 MSI 图形安装

MSI 就是 Microsoft Installer 的简写，是微软格式的安装包，也是 Windows 为自己用户做的可扩展软件按管理系统。它用来管理软件的安装、组件添加和删除、监视文件修复及回滚。它实际上是一个数据库，包含安装产品所需要的信息和在复杂情况下进行安装、卸载程序所需的指令和数据，就是微软为 Windows 弄的一个安装程序模块，软件包上它（很多软件直接就是 msi 了），用户就可以直接 next、next、next⋯finish 等安装软件。

优点：简单、方便、快捷，安装配置都是用默认值。

缺点：缺点就是安装过程都是默认设计的，无法直接进行个性化设计、微调，没办法专项安装数据库。

3.1.1 MSI 图形安装包下载

MSI 图形安装包下载界面如图 3-1 所示。

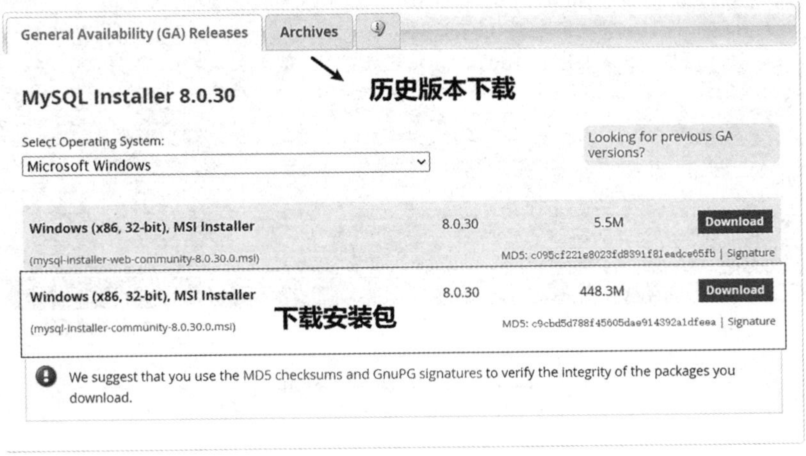

图 3-1 MSI 图形安装包下载界面

3.1.2 系统环境要求

(1)部分操作系统需要更新证书,并启动 Windows Installer 服务。
如果安装过程有对环境依赖的报错提示,请按以下顺序安装修复,需重启生效。

a. NET4.5 环境:安装 NDP452 - KB2901907 - x86 - x64 - AllOS - ENU. exe
b. c + + 2019 运行库:安装 MSVBCRT. AIO_v2020.05.20 微软常用运行库 64 位 . exe
c. 操作系统打补丁:VC_redist. x86. exe

说明:企业服务器中常出现此问题,需要根据不同的操作系统下载对应的版本,而且部分操作系统还需要打补丁,具体可参考操作系统官网说明。

3.1.3 安装过程

MSI 安装过程如图 3-2~图 3-11 所示。
注意事项:
(1)通过此方法设置环境变量 setx path "E:\Program Files\MySQL\MySQL Server 8.0\bin"。
(2)安装过程如果启动不了,删除 MySQL 服务,设置下组权限,重启后,要重新安装。
(3)删除服务,请使用:sc delete 服务名。
(4)启动关闭服务:net start/stop MYSQL80。
(5)可以将其写到一个 . bat 里边,方便运行:@cmd /k net start MYSQL80。

图 3-2 MSI 安装初始化

第 3 章　MySQL 安装及卸载

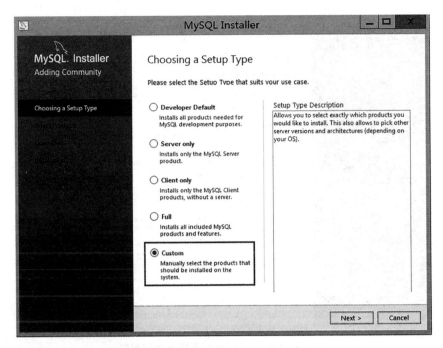

图 3-3　MSI 安装自定义类别

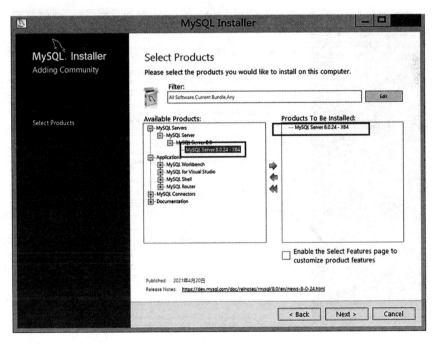

图 3-4　MSI 安装的服务及目录

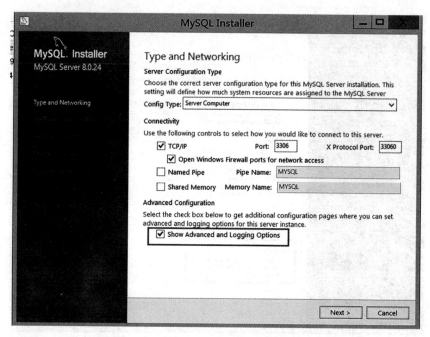

图 3-5　MSI 安装的端口及服务类型配置

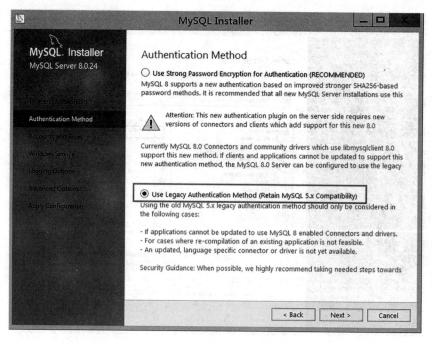

图 3-6　MSI 安装弱密码选项

第 3 章　MySQL 安装及卸载

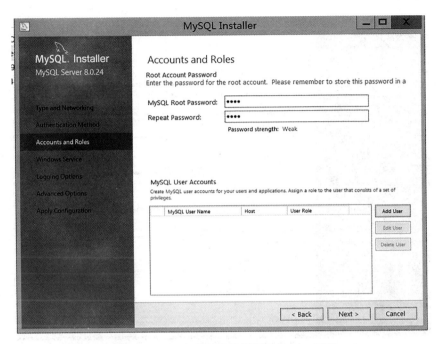

图 3-7　MSI 安装设定超级管理员 root 密码

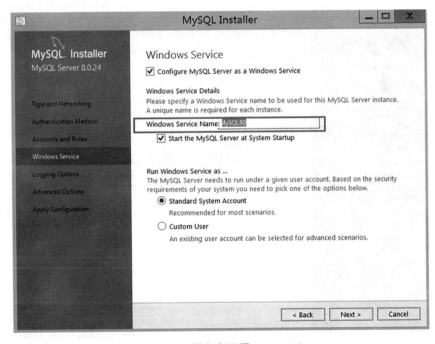

图 3-8　服务名配置 MYSQL80

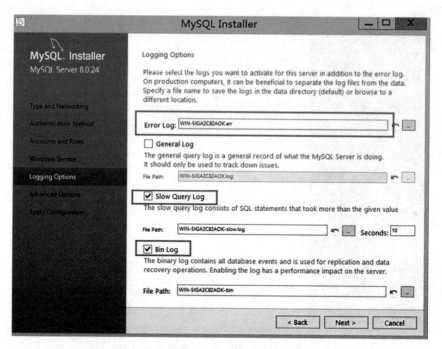

图 3-9　日志文件配置

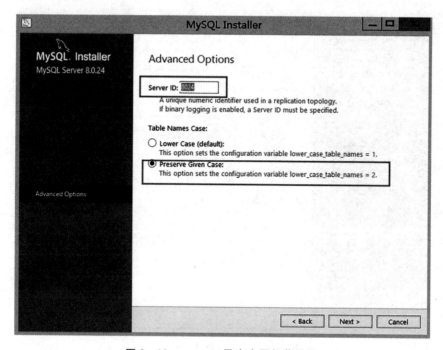

图 3-10　serverid 及大小写规范设置

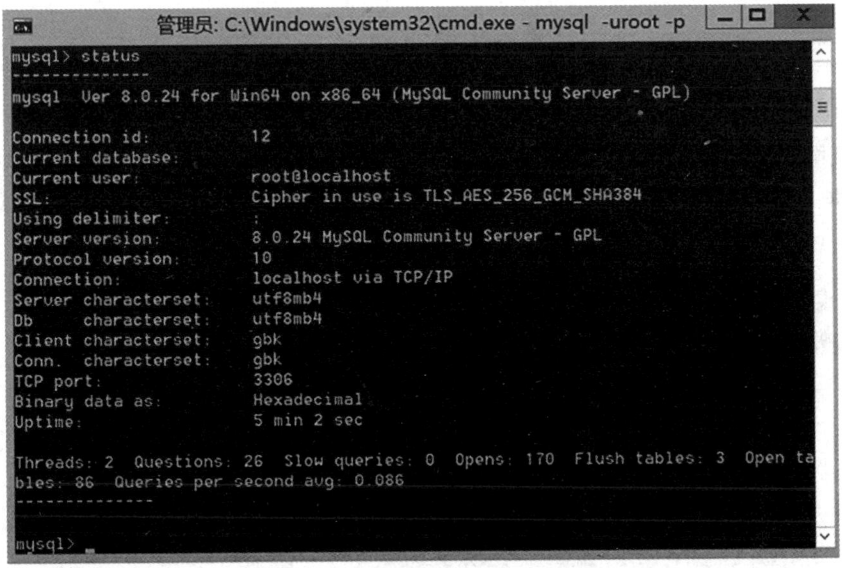

图 3-11　MySQL8 登录界面信息

3.2　Windows 环境二进制安装

使用二进制包安装 MySQL,也称静默安装,优点是简单,好维护,大部分环境都适用(推荐),如图 3-12 所示。

图 3-12　Windows 二进制安装包下载界面

3.2.1 Window 环境二进制安装过程

1. 初始化

E:\mysql-8.0.26-winx64\bin\mysqld --initialize --user=mysql --console --basedir=E:\mysql-8.0.26-winx64 --datadir=E:\mysql-8.0.26-winx64\data80263307

注意：全路径安装，命令放在同一行，空格以及 cmd 窗口用管理员权限执行。

说明：

--initialize 表示随机密码，如果加-insecure 表示空密码

--console 表示 cmd 窗口打印日志

--basedir 表示 mysql 安装的根目录

--datadir 表示 mysql 安装的数据文件目录

MySQL 二进制安装包初始化 cmd 界面如图 3-13 所示。

图 3-13　MySQL 二进制安装包初始化 cmd 界面

2. 参数文件配置

[mysqld]

(1)设置 3307 端口，二进制可以在一个服务器安装多个，故需设置端口来区分

port=3307

(2)设置 MySQL 的安装目录

basedir=E:\mysql-8.0.26-winx64

(3)设置 MySQL 数据库的数据的存放目录

datadir=E:\mysql-8.0.26-winx64\data80263307

(4)允许最大连接数

max_connections=200

(5)允许连接失败的次数。这是为了防止有人从该主机试图攻击数据库系统

max_connect_errors=10

(6)服务端使用的字符集默认为 utf8mb4

character-set-server=utf8mb4

(7)创建新表时将使用的默认存储引擎

default-storage-engine=INNODB

(8)MySQL8 默认使用"mysql_native_password"插件认证

default_authentication_plugin=mysql_native_password

3. mysql 实例安装(图 3-14)

E:\mysql-8.0.26-winx64\bin\mysqld install mysql80263307 --defaults-file="E:\mysql-8.0.26-winx64\data80263307\mysql80263307.ini"

参数说明：

mysql80263307:表示 MySQL 服务名

defaults-file:指定配置好的参数文件

删除服务:sc delete mysql80263307

启动服务 net start mysql80263307

安装好之后,MySQL8 需要重置密码才可以正常使用。

```
C:\Users\Administrator>E:\mysql-8.0.26-winx64\bin\mysqld install mysql80263307 --defaults-file="E:\mysql-8.0.26-winx64\data80263307\mysql80263307.ini"
Service successfully installed.

C:\Users\Administrator>
```

图 3-14 MySQL 二进制实例安装 cmd 界面

3.2.2 重置密码

MySQL8 安装好后需要创建远程 root 用户,重置密码才可以正常使用,按以下步骤重置。

1. 先关闭 MySQL 服务

net stop mysql80263307

其中 mysql80263307 表示 MySQL 服务名

2. 跳过密码验证启动

E:\mysql-8.0.26-winx64\bin\mysqld
--datadir=E:\mysql-8.0.26-winx64\data80263307
--console --skip-grant-tables --shared-memory

参数说明：

console 表示打印日志

skip-grant-tables 表示登录时跳过权限检查

shared-memory 表示共享内存的连接方式

3. 重新登录 MySQL

创建远程 root 用户及重建密码。

cd G:\mysql-8.0.23-winx64\bin

mysql-uroot-p-P3307

Mysql> flush privileges;

Mysql> update mysql.user set authentication_string=password('root') where user='root';

Mysql> create user root@'%' identified with mysql_native_password by 'root';

Mysql> grant all on *.* to root@'%' with grant option;

Mysql> flush privileges;

注意：如果出现提示密码验证不符合密码策略，可参考后面密码重置章节。

4. 重新登录 MySQL

cd G:\mysql-8.0.23-winx64\bin

mysql-uroot-p-P3307

3.3 Linux PRM 包安装 MySQL

RPM 是 Redhat Package Manager 的缩写，是由 RedHat 公司开发的软件包安装和管理程序。使用 RPM 用户可以自行安装和管理 Linux 上的应用程序和系统工具。可以让用户直接以 binary 方式安装软件包，用此种方式安装 MySQL，方便快捷，实践中推荐使用此种方法。以下部署是用 RPM 包基于 RHEL8 的操作系统安装 MySQL8.0。

3.3.1 环境准备

1. 操作系统

[root@jeames ~]# cat /etc/redhat-release

Red Hat Enterprise Linux release 8.1 (Ootpa)

2. 关闭防火墙

(1) 查看防火墙状态

[root@jeames ~]# systemctl status firewalld

(2) 关闭防火墙

[root@jeames ~]# systemctl stop firewalld

(3) 取消开机自启动

[root@jeames ~]# systemctl disable firewalld
Removed /etc/systemd/system/multi – user. target. wants/firewalld. service
Removed /etc/systemd/system/dbus – org. fedoraproject. FirewallD1. service

3. selinux 关闭

修改参数文件/etc/sysconfig/selinux 中 SELINUX 的值为 disabled

[root@jeames ~]# sed – i 's/SELINUX\ = enforcing/SELINUX\ = disabled/g' /etc/selinux/contig

修改完成后需要重启才生效。

3.3.2 RPM 安装包下载

官网下载地址:https://dev. mysql. com/downloads/mysql/。

RPM 包下载选择 Red Hat Enterprise Linux / Oracle Linux,界面如图 3 – 15 所示。

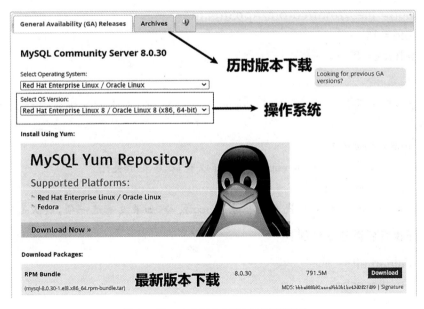

图 3 – 15 MySQL8 RPM 包下载界面

3.3.3 离线 yum 配置

这里配置本地 yum 源,企业服务器不连外网,一般使用插入镜像盘或者 U 盘做依赖源,yum 配置好后安装 MySQL 的依赖包即可。

1. 创建挂载路径

mkdir -p /mnt/cdrom

2. 挂载系统镜像光盘到指定目录

mount -t iso9660 /dev/sr0 /mnt/cdrom

3. 修改 yum 源配置文件

(1) 编辑 rhel8-local.repo 文件
cd /etc/yum.repos.d
vi rhel8-local.repo

[localREPO]
name=localhost8
baseurl=file:///mnt/cdrom/BaseOS
enable=1
gpgcheck=0

[localREPO_APP]
name=localhost8_app
baseurl=file:///mnt/cdrom/AppStream
enable=1
gpgcheck=0

(2) baseurl

这是非常重要的节点,表明了 repositry 的地址,支持 ftp 协议,http 协议和 file 协议。

(3) enabled=0/1
只有两个值,为 1 时表示 repositry 可以获取,0 表示关闭。

(4) gpgcheck=0/1
有 1 和 0 两个选择,分别代表是否进行 gpg 校验,如果没有这一项,默认是检查的。

4. 配置好后重建本地缓存

yum clean all
yum makecache
yum repolist

这一步主要是确保 yum 能正常使用。

5. 安装依赖包

yum - y install lrzsz wget perl - Digest - MD5
yum - y install numactl
yum - y install ncurses ncurses - devel openssl - devel bison gcc gcc - c + + make

这些依赖包都是 MySQL 8.0 版本所必要的。

3.3.4 添加用户及组

以 root 用户登录到服务器,在创建用户及组之前,检查是否存在 mysql 用户组。

1. 添加组

[root@jeames ~]# groupadd mysql

删除组:groupdel users

2. 用户加到组,并指定根目录

[root@jeames ~]# useradd - g mysql - d /home/mysql mysql

删除用户:userdel mysql

3. 修改密码

[root@jeames ~]# passwd mysql
New password:
BAD PASSWORD: The password is shorter than 8 characters
Retype new password:
passwd: all authentication tokens updated successfully

3.3.5 安装 MySQL

将下载好的 rpm 包上传到服务器,按照如下步骤安装部署。

1. 校验安装包

[root@jeames ~]# md5sum mysql - 8.0.27 - 1.el8.x86_64.rpm - bundle.tar
edf4d0f95867f62cdcc97b7349bedc59 mysql - 8.0.27 - 1.el8.x86_64.rpm - bundle.tar

md5sum 是 linux 下的 shell 命令,功能是计算检验 MD5 校验码。

2. 解压安装包

[root@jeames ~]# tar - xvf mysql - 8.0.27 - 1.el8.x86_64.rpm - bundle.tar -C /home/mysql/

必须安装的 4 个 rpm 包：

mysql-community-common-8.0.27-1.el8.x86_64.rpm --必须，它是公共库

mysql-community-libs-8.0.27-1.el8.x86_64.rpm --它是开发库

mysql-community-client-8.0.27-1.el8.x86_64.rpm --它是客户端的安装包

mysql-community-server-8.0.27-1.el8.x86_64.rpm --它是服务端的安装包

3. 授权

[root@jeames ~]# chown -R mysql:mysql /home/mysql/

4. rpm 包安装

[root@jeames ~]# cd /home/mysql/

注意：包之间相互依赖，所以必须注意安装顺序，先装 common，接着装 libs，然后再装 client，最后装 server。

[root@jeames mysql]# rpm -ivh mysql-community-common-8.0.27-1.el8.x86_64.rpm

[root@jeames mysql]# rpm -ivh --force --nodeps mysql-community-libs-8.0.27-1.el8.x86_64.rpm

[root@jeames mysql]# rpm -ivh mysql-community-client-8.0.27-1.el8.x86_64.rpm --force --nodeps

[root@jeames mysql]# rpm -ivh mysql-community-server-8.0.27-1.el8.x86_64.rpm --force --nodeps

5. 确认 rpm 包安装是否已安装

[root@jeames mysql]# rpm -qa | grep mysql

mysql-community-common-8.0.27-1.el8.x86_64

mysql-community-server-8.0.27-1.el8.x86_64

mysql-community-libs-8.0.27-1.el8.x86_64

mysql-community-client-8.0.27-1.el8.x86_64

6. 初始化

[root@jeames mysql]# mysqld --initialize

7. 大小写铭感参数设置

cat /etc/my.cnf 配置文件里可以修改

只需要在[mysqld]下面添加一行配置，即 lower_case_table_names = 0

其中 0 表示区分大小写，1 表示不区分大小写

此处的目的是开发的规范,Linux 环境一般要求区分大小写。

3.3.6 MySQL 启动

使用 rpm 包安装,在 RHEL8 操作系统可使用 systemctl 方便灵活地管理 MySQL 服务,注意记得设置开机自启动。

1. 查看 MySQL 服务

[root@jeames ~]# ps – ef | grep mysql
[root@jeames ~]# systemctl status mysqld

2. 文件所有者和文件关联组授权

[root@jeames ~]# chown mysql:mysql /var/lib/mysql – R

3. 启动服务

[root@jeames ~]# systemctl start mysqld
[root@jeames ~]# systemctl status mysqld

4. 设置开机自启动

[root@jeames ~]# systemctl enable mysqld

5. 查看临时密码

[root@jeames ~]# cat /var/log/mysqld.log | grep password

如图 3-16 所示,确认 MySQL 的服务已正常运行,若无法启动,请排查错误日志。

```
[root@jeames mysql]# systemctl status mysqld
● mysqld.service - MySQL Server
   Loaded: loaded (/usr/lib/systemd/system/mysqld.service; enabled; vendor preset: disabled)
   Active: active (running) since Sat 2022-03-26 14:57:32 CST; 6s ago
     Docs: man:mysqld(8)
           http://dev.mysql.com/doc/refman/en/using-systemd.html
  Process: 37690 ExecStartPre=/usr/bin/mysqld_pre_systemd (code=exited, status=0/SUCCESS)
 Main PID: 37718 (mysqld)
   Status: "Server is operational"
    Tasks: 38 (limit: 11336)
   Memory: 432.7M
   CGroup: /system.slice/mysqld.service
           └─37718 /usr/sbin/mysqld

Mar 26 14:57:28 jeames systemd[1]: Starting MySQL Server...
Mar 26 14:57:32 jeames systemd[1]: Started MySQL Server.
```

图 3-16 MySQL 服务状态

3.3.7 创建远程用户

MySQL 的用户认证形式是:用户名+主机,MySQL 的主机信息可以是本地 (localhost),某个 IP,某个 IP 段,以及任何地方等。如要从其他 IP 访问 MySQL,就需要创建远程用户。

1. 用临时密码登录

[root@jeames ~]#　grep 'temporary password' /var/log/mysqld.log
[root@jeames ~]#　mysql – uroot – p

2. 修改本地 root 用户密码

mysql> alter user root@'localhost' identified with mysql_native_password by 'root'

mysql> flush privileges

mysql> select Host,User from mysql.user

3. 创建远程 root 用户

mysql> create user root@'%' identified with mysql_native_password by 'root'

mysql> grant all on *.* to root@'%' with grant option

mysql> flush privileges

修改后,通过其他 IP 就可以正常访问 MySQL 数据库了,如图 3-17 所示。

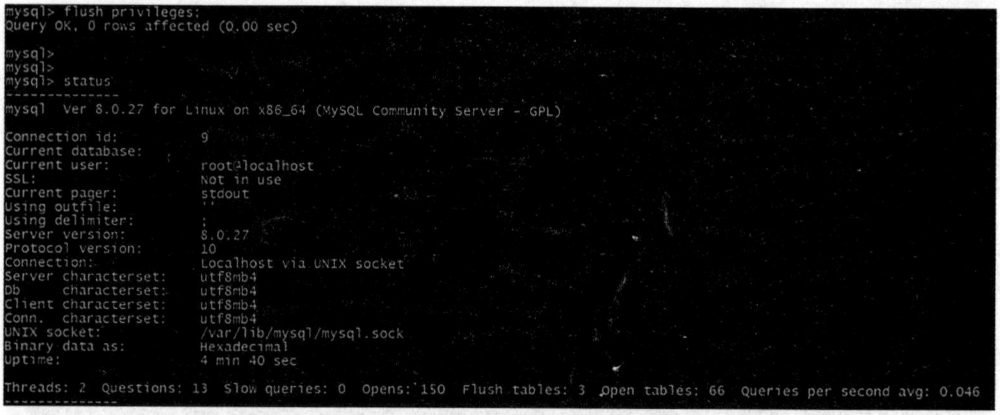

图 3-17　MySQL 登录后的数据库信息

3.4　Linux 二进制多版本部署 MySQL

　　Linux MySQL 二进制安装包已经过编译,安装简单,可以安装到任何路径下,灵活性好,一台服务器可以安装多个 MySQL,本次部署 5.6/5.7/8.0 三个版本给大家展示。

3.4.1 环境准备

安装数据库之前,操作系统环境至关重要,本次部署是基于 RHEL8 操作系统。

1. 操作系统

[root@jeames ~]#　cat /etc/redhat‐release
Red Hat Enterprise Linux release 8.1 (Ootpa)

2. 关闭防火墙

(1) 查看防火墙状态
[root@jeames ~]#　systemctl status firewalld
(2) 关闭防火墙
[root@jeames ~]#　systemctl stop firewalld
(3) 取消开机自启动
[root@jeames ~]#　systemctl disable firewalld
Removed/etc/systemd/system/multi‐user.target.wants/firewalld.service.
Removed/etc/systemd/system/dbus‐org.fedoraproject.FirewallD1.service.

3. selinux 关闭

修改参数文件/etc/sysconfig/selinux 中 SELINUX 的值为 disabled
[root@jeames ~]#　sed ‐i 's/SELINUX\= enforcing/SELINUX\= disabled/g' /etc/selinux/config
修改完成后需要重启服务器才生效。

3.4.2 安装包下载

Linux‐Generic 下载界面如图 3‐18 所示。

图 3‐18　MySQL Linux‐Generic 下载界面

3.4.3 配置 yum 安装依赖

这里配置本地 yum 源,企业服务器不连外网,一般使用插入镜像盘或者 U 盘做依赖源,yum 配置好后安装 MySQL 的依赖包即可。

1. 创建挂载路径

mkdir -p /mnt/cdrom

2. 挂载系统镜像光盘到指定目录

#因为光盘的格式通常是 iso9660,意思是/dev/sr0 挂载在/mnt/cdrom 目录上
mount -t iso9660 /dev/sr0 /mnt/cdrom

3. 修改 yum 源配置文件

##编辑 rhel8-local.repo 文件,加入以下内容
[root@jeames ~]# cd /etc/yum.repos.d
[root@jeames yum.repos.d]# vi rhel8-local.repo

[localREPO]
name=localhost8
baseurl=file:///mnt/cdrom/BaseOS
enable=1
gpgcheck=0

[localREPO_APP]
name=localhost8_app
baseurl=file:///mnt/cdrom/AppStream
enable=1
gpgcheck=0

4. 配置好后,重建本地缓存

yum clean all
yum makecache
yum repolist

5. 安装 MySQL 8.0 版本二进制所需的依赖包

yum -y install libncurses*
yum -y install libaio
yum -y install perl perl-devel
yum -y install autoconf

```
yum -y install numactl.x86_64
```

##通过 rpm -qa 可以查询是否将依赖包安装成功

```
[root@jeames yum.repos.d]# rpm -qa libaio perl perl-devel autoconf \
numactl.x86_64libncurses
```

libaio-0.3.112-1.el8.x86_64
perl-5.26.3-416.el8.x86_64
perl-devel-5.26.3-416.el8.x86_64
autoconf-2.69-27.el8.noarch

3.4.4 安装包解压缩

MySQL 不同版本的后缀格式各不相同,在 Linux 环境下需要通过不同的命令解压。

1. 创建安装目录

[root@jeames ~]# mkdir /soft
[root@jeames ~]# mkdir /usr/local/mysqlsoft

2. 上传软件包

通过 Xftp 工具或者 SecureCRT 自带的 Sftp 上传软件包
mysql-5.6.48-linux-glibc2.12-x86_64.tar
mysql-5.7.30-linux-glibc2.12-x86_64.tar.gz
mysql-8.0.19-linux-glibc2.12-x86_64.tar

3. 解压缩

[root@jeames ~]# cd /soft/
[root@jeames soft]# tar -zxvf mysql-5.6.48-linux-glibc2.12-x86_64.tar.gz -C /usr/local/mysqlsoft
[root@jeames soft]# tar -zxvf mysql-5.7.30-linux-glibc2.12-x86_64.tar.gz -C /usr/local/mysqlsoft
[root@jeames soft]# tar -Jxf mysql-8.0.19-linux-glibc2.12-x86_64.tar.xz -C /usr/local/mysqlsoft

4. 快捷方式创建

[root@jeames ~]# mkdir -p /usr/local/mysql56
[root@jeames ~]# mkdir -p /usr/local/mysql57
[root@jeames ~]# mkdir -p /usr/local/mysql80

通过以下方式创建软连接,主要是方便管理

[root@jeames ~]# ln -s /usr/local/mysqlsoft/mysql-5.6.48-linux-glibc2.12-x86_64 /usr/local/mysql56/mysql5648

[root@jeames ~]# ln -s /usr/local/mysqlsoft/mysql-5.7.30-linux-glibc2.12-x86_64 /usr/local/mysql57/mysql5730

[root@jeames ~]# ln -s /usr/local/mysqlsoft/mysql-8.0.19-linux-glibc2.12-x86_64 /usr/local/mysql80/mysql8019

[root@jeames ~]# cd /usr/local
[root@jeames local]# ll
total 0
drwxr-xr-x. 2 root root   6 Aug 12 2018 bin
drwxr-xr-x. 2 root root   6 Aug 12 2018 etc
drwxr-xr-x. 2 root root   6 Aug 12 2018 games
drwxr-xr-x. 2 root root   6 Aug 12 2018 include
drwxr-xr-x. 2 root root   6 Aug 12 2018 lib
drwxr-xr-x. 2 root root   6 Aug 12 2018 lib64
drwxr-xr-x. 2 root root   6 Aug 12 2018 libexec
drwxr-xr-x. 2 root root   23 Aug 21 00:01 mysql56
drwxr-xr-x. 2 root root   23 Aug 21 00:01 mysql57
drwxr-xr-x. 2 root root   23 Aug 21 00:01 mysql80
drwxr-xr-x. 5 root root 135 Aug 20 23:56 mysqlsoft
drwxr-xr-x. 2 root root   6 Aug 12 2018 sbin
drwxr-xr-x. 5 root root   49 Jul 26 2021 share
drwxr-xr-x. 2 root root   6 Aug 12 2018 src

软连接是 Linux 中一个常用命令,它的功能是为某个文件在另外一个位置建立一个同步的链接。简单来说,就是 Windows 里面的快捷方式。图 3-19 所示为 MySQL 二进制包解压后的创建的软连接。

```
lrwxrwxrwx. 1 root root 56 Aug 21 00:01 mysql5648 -> /usr/local/mysqlsoft/mysql-5.6.48-linux-glibc2.12-x86_64
[root@jeames mysql56]# cd ..
[root@jeames local]# cd mysql57
[root@jeames mysql57]# ll
total 0
lrwxrwxrwx. 1 root root 56 Aug 21 00:01 mysql5730 -> /usr/local/mysqlsoft/mysql-5.7.30-linux-glibc2.12-x86_64
[root@jeames mysql57]# cd ..
[root@jeames local]# cd mysql80
[root@jeames mysql80]# ll
total 0
lrwxrwxrwx. 1 root root 56 Aug 21 00:01 mysql8019 -> /usr/local/mysqlsoft/mysql-8.0.19-linux-glibc2.12-x86_64
```

图 3-19 MySQL 二进制包解压后的软连接

3.4.5 组用户及授权

使用 root 账户登录 RHEL8 操作系统的服务器,然后执行如下命令,添加 mysql 用户组及用户,更改目录权限,如图 3-20 所示。

1. 新增 mysql 用户组

[root@jeames ~]# groupadd mysql

2. 添加 mysql 用户

[root@jeames ~]# useradd -r -g mysql mysql

注意:useradd 命令用于建立用户账号,-r 表示建立系统账号,-g 指定用户所属的群组

3. 安装包指定用户和组

[root@jeames ~]# chown -R mysql. mysql /usr/local/mysqlsoft/

chown 将指定文件的拥有者改为指定的用户或组,-R 表示级联。

```
[root@jeames ~]# groupadd mysql
[root@jeames ~]# useradd -r -g mysql mysql
[root@jeames ~]# chown -R mysql.mysql /usr/local/mysqlsoft/
[root@jeames ~]# cd /usr/local/mysqlsoft/
[root@jeames mysqlsoft]# ll -rt
total 0
drwxr-xr-x. 13 mysql mysql 191 Aug 20 23:55 mysql-5.6.48-linux-glibc2.12-x86_64
drwxr-xr-x.  9 mysql mysql 129 Aug 20 23:56 mysql-5.7.30-linux-glibc2.12-x86_64
drwxr-xr-x.  9 mysql mysql 129 Aug 20 23:57 mysql-8.0.19-linux-glibc2.12-x86_64
```

图 3-20 MySQL 二进制包指定的用户及组

3.4.6 二进制安装

进入数据库目录,所有配置都在软连接目录下。在 root 用户下通过以下命令完成 MySQL 初始化。

1. 5.6 版本安装(图 3-21)

/usr/local/mysql56/mysql5648/scripts/mysql_install_db --user= mysql \
-- basedir= /usr/local/mysql56/mysql5648
-- datadir= /usr/local/mysql56/mysql5648/data

2. 5.7 版本安装(图 3-22)

/usr/local/mysql57/mysql5730/bin/mysqld -- initialize - insecure -- user = mysql \
-- basedir= /usr/local/mysql57/mysql5730
-- datadir= /usr/local/mysql57/mysql5730/data

3. 8.0 版本安装(图 3-23)

/usr/local/mysql80/mysql8019/bin/mysqld －－initialize－insecure －－user=mysql \

－－basedir=/usr/local/mysql80/mysql8019

－－datadir=/usr/local/mysql80/mysql8019/data

#－－basedir 表示 mysql 根目录

#－－datadir 表示 mysql 数据文件目录

#－－initialize-insecure 表示空密码

```
To start mysqld at boot time you have to copy
support-files/mysql.server to the right place for your system

PLEASE REMEMBER TO SET A PASSWORD FOR THE MySQL root USER !
To do so, start the server, then issue the following commands:

 /usr/local/mysql56/mysql5648/bin/mysqladmin -u root password 'new-password'
 /usr/local/mysql56/mysql5648/bin/mysqladmin -u root -h jeames password 'new-password'

Alternatively you can run:

 /usr/local/mysql56/mysql5648/bin/mysql_secure_installation

which will also give you the option of removing the test
databases and anonymous user created by default.  This is
strongly recommended for production servers.

See the manual for more instructions.

You can start the MySQL daemon with:

 cd . ; /usr/local/mysql56/mysql5648/bin/mysqld_safe &

You can test the MySQL daemon with mysql-test-run.pl

 cd mysql-test ; perl mysql-test-run.pl

Please report any problems at http://bugs.mysql.com/

The latest information about MySQL is available on the web at

 http://www.mysql.com

Support MySQL by buying support/licenses at http://shop.mysql.com

New default config file was created as /usr/local/mysql56/mysql5648/my.cnf and
will be used by default by the server when you start it.
You may edit this file to change server settings
```

图 3-21 MySQL 5.6 版本安装成功界面

```
[root@jeames ~]# /usr/local/mysql57/mysql5730/bin/mysqld --initialize-insecure --user=mysql \
> --basedir=/usr/local/mysql57/mysql5730 --datadir=/usr/local/mysql57/mysql5730/data
2022-08-20T16:44:44.920529Z 0 [Warning] TIMESTAMP with implicit DEFAULT value is deprecated. Please use --explicit_defaults_for_timestamp server option (see documentation for more details).
2022-08-20T16:44:45.234416Z 0 [Warning] InnoDB: New log files created, LSN=45790
2022-08-20T16:44:45.301729Z 0 [Warning] InnoDB: Creating foreign key constraint system tables.
2022-08-20T16:44:45.365623Z 0 [Warning] No existing UUID has been found, so we assume that this is the first time that this server has been started. Generating a new UUID: 658c0640-20a7-11ed-bd62-000c299e9213.
2022-08-20T16:44:46.366925Z 0 [Warning] Gtid table is not ready to be used. Table 'mysql.gtid_executed' cannot be opened.
2022-08-20T16:44:46.217332Z 0 [Warning] CA certificate ca.pem is self signed.
2022-08-20T16:44:46.430167Z 1 [Warning] root@localhost is created with an empty password ! Please consider switching off the --initialize-insecure option.
```

图 3-22 MySQL 5.7 版本安装成功界面

```
[root@jeames ~]# /usr/local/mysql80/mysql8019/bin/mysqld --initialize-insecure --user=mysql \
> --basedir=/usr/local/mysql80/mysql8019 --datadir=/usr/local/mysql80/mysql8019/data
2022-08-20T16:45:51.442274Z 0 [System] [MY-013169] [Server] /usr/local/mysql80/mysql8019/bin/mysqld (mysqld 8.0.19) initializing of server in progress as process 34370
2022-08-20T16:45:54.196567Z 5 [Warning] [MY-010453] [Server] root@localhost is created with an empty password ! Please consider switching off the --initialize-insecure option.
```

图 3-23　MySQL 8.0 版本安装成功界面

3.4.7　管理多实例

创建对应目录,放置错误日志,以便于后期排查相关错误。

如图 3-24 所示。mysqld_multi 旨在管理多个 mysqld 进程,这些进程可根据不同 Unix 套接字文件和 TCP/IP 端口上的链接。它可以启动或停止服务器,或报告其当前状态,通过编辑参数文件来实现。

[root@jeames ~]#　vi /etc/my.cnf
[mysqld_multi]
mysqld= /usr/local/mysql80/mysql8019/bin/mysqld_safe
mysqladmin= /usr/local/mysql80/mysql8019/bin/mysqladmin
log= /usr/local/mysqlsoft/log/mysqld_multi.log
user= root
password= root

[mysql]
default-character-set= utf8mb4

[mysqld56483507]
mysqld= /usr/local/mysql56/mysql5648/bin/mysqld_safe
mysqladmin= /usr/local/mysql56/mysql5648/bin/mysqladmin
port= 3507
basedir= /usr/local/mysql56/mysql5648
datadir= /usr/local/mysql56/mysql5648/data
socket= /usr/local/mysql56/mysql5648/data/mysqls56483507.sock
server_id= 56483507
log-bin
default-time-zone = '+ 8:00'
skip-name-resolve
character_set_server= utf8mb4

[mysqld57303508]
mysqld= /usr/local/mysql57/mysql5730/bin/mysqld_safe
mysqladmin= /usr/local/mysql57/mysql5730/bin/mysqladmin

port=3508
basedir=/usr/local/mysql57/mysql5730
datadir=/usr/local/mysql57/mysql5730/data
socket=/usr/local/mysql57/mysql5730/data/mysqls57303508.sock
server_id=57303508
log-bin
default-time-zone='+8:00'
log_timestamps=SYSTEM
skip-name-resolve
character_set_server=utf8mb4

[mysqld80193509]
mysqld=/usr/local/mysql80/mysql8019/bin/mysqld_safe
mysqladmin=/usr/local/mysql80/mysql8019/bin/mysqladmin
port=3509
basedir=/usr/local/mysql80/mysql8019
datadir=/usr/local/mysql80/mysql8019/data
socket=/usr/local/mysql80/mysql8019/data/mysqls80193509.sock
default_authentication_plugin=mysql_native_password
server_id=80193509
log-bin
default-time-zone='+8:00'
log_timestamps=SYSTEM
skip-name-resolve
character_set_server=utf8mb4

(1) 环境变量创建

[root@jeames ~]# echo "export PATH=$PATH:/usr/local/mysql80/mysql8019/bin" >> /root/.bashrc
[root@jeames ~]# source /root/.bashrc
[root@jeames ~]# mysqld_multi report
Reporting MySQL servers
MySQL server from group: mysqld56483507 is not running
MySQL server from group: mysqld57303508 is not running
MySQL server from group: mysqld80193509 is not running

(2) mysqld_multi 命令管理

启动全部实例：mysqld_multi start

关闭全部实例:mysqld_multi stop

[root@jeames ~]#　mysqld_multi start
[root@jeames ~]#　mysqld_multi report
Reporting MySQL servers
MySQL server from group: mysqld56483507 is running
MySQL server from group: mysqld57303508 is running
MySQL server from group: mysqld80193509 is running

注意：mysqld_multi 不能关闭实例,请按照如下方法处理,需重置数据库密码为 root。

可以通过 sock 方式登录多版本

[root@jeames ~]#　mysql-uroot-p-S /usr/local/mysql80/mysql8019/data/mysqls80193509.sock
set password for root@'localhost'=password('root');　--5.6 版本
update mysql.user set authentication_string=password('root') where user='root';--5.7 版本
ALTER USER 'root'@'localhost' IDENTIFIED WITH MYSQL_NATIVE_PASSWORD BY'root';--8.0 版本

[root@jeames ~]#　which mysqld_multi
/usr/local/mysql80/mysql8019/bin/mysqld_multi

需要修改 mysqld_multi 的 221 行

my $com = join '', 'my_print_defaults ', @defaults_options, $group;

替换为：

my $com = join '', 'my_print_defaults-s', @defaults_options, $group;

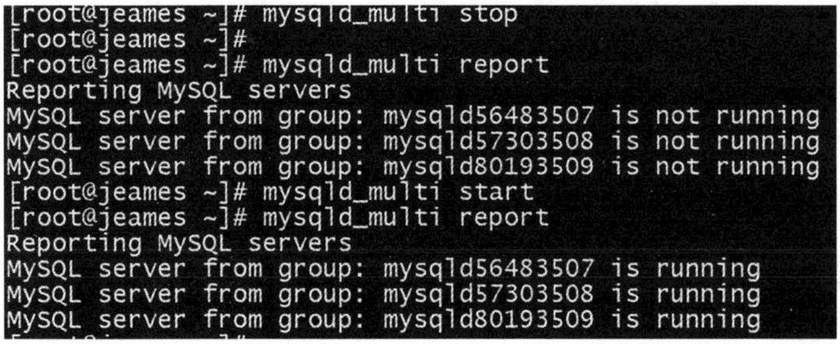

图 3-24　MySQL 8.0 mysqld_multi 管理

3.5 Linux 源码安装 MySQL

Linux 源码安装 MySQL 应在实际安装的操作系统中，根据需要定制编译，此方法灵活性好。一台服务器可以安装多个 MySQL，但安装过程较复杂，部分依赖包需要网上下载，编译时间长。

3.5.1 源码安装包下载

官网下载地址：https://dev.mysql.com/downloads/mysql/

源码包下载选择 Source Code，无须选择操作系统平台，界面如图 3-25 所示。

图 3-25　MySQL 8.0 源码安装包下载界面

3.5.2 环境准备

源码安装对环境要求比较高，涉及部分系统内核的升级。

1. 操作系统

[root@jeames ~]# cat /etc/redhat-release
Red Hat Enterprise Linux release 8.1 (Ootpa)

2. 关闭防火墙

#查看防火墙状态
[root@jeames ~]# systemctl status firewalld
#关闭防火墙

[root@jeames ~]#　systemctl stop firewalld
取消开机自启动
[root@jeames ~]#　systemctl disable firewalld
Removed /etc/systemd/system/multi-user.target.wants/firewalld.service.
Removed /etc/systemd/system/dbus-org.fedoraproject.FirewallD1.service.

3. selinux 关闭

修改参数文件/etc/sysconfig/selinux 中 SELINUX 的值为 disabled

[root@jeames ~]#　sed -i 's/SELINUX\=enforcing/SELINUX\=disabled/g' /etc/selinux/config

修改完成后需要重启服务器才生效。

4. Host 解析配置

[root@jeames ~]#　hostname
jeames
[root@jeames ~]#　vi /etc/hosts
127.0.0.1 localhost localhost.localdomain localhost4 localhost4.localdomain4
::1 localhost localhost.localdomain localhost6 localhost6.localdomain6
此处新增 hostname 与 ip 的对应
192.168.1.88 jeames

5. 卸载 mariadb

为了保证后续操作不会产生其他冲突，我们卸载部分操作系统自带原有的 mariadb

[root@jeames ~]#　rpm -qa| grep mariadb
[root@jeames ~]#　rpm -qa| grep mariadb-libs| xargs rpm -e --nodeps

3.5.3　配置 yum 安装依赖

这里需配置本地 yum 源，可以复制镜像文件，挂载目的方式作为 yum 源，yum 配置好后一定要安装 MySQL 的依赖包，不然源码编译会失败。

1. 创建挂载路径

[root@jeames ~]#　mkdir -p /mnt/cdrom

2. 挂载系统镜像光盘到指定目录

因为光盘的格式通常是 iso9660,意思是/dev/sr0 挂载在/mnt/cdrom 目录上
[root@jeames ~]#　mount -t iso9660 /dev/sr0 /mnt/cdrom

mount: /mnt/cdrom: WARNING: device write-protected, mounted read-only.

3. 修改 yum 源配置文件

编辑 rhel8-local.repo 文件

[root@jeames ~]# cd /etc/yum.repos.d
[root@jeames yum.repos.d]# vi rhel8-local.repo
[localREPO]
name=localhost8
baseurl=file:///mnt/cdrom/BaseOS
enable=1
gpgcheck=0

[localREPO_APP]
name=localhost8_app
baseurl=file:///mnt/cdrom/AppStream
enable=1
gpgcheck=0

4. 配置好后重建本地缓存

yum clean all
yum makecache
yum repolist

5. 安装 MySQL 8.0 源码所需的依赖包

[root@jeames ~]# yum -y install lrzsz wget perl-Digest-MD5
[root@jeames ~]# yum -y install libaio

编译软件依赖

[root@jeames ~]# yum -y install cmake gcc gcc-c++
[root@jeames ~]# yum -y install git make
[root@jeames ~]# yum install libtirpc-devel
[root@jeames ~]# yum insatll rpcgen
[root@jeames ~]# yum install libudev-devel
[root@jeames ~]# yum install ncurses-devel

字符终端处理依赖

[root@jeames ~]# yum install -y openssl openssl-devel ncurses
注意：rpcgen 需要手动下载安装包编译安装。
rpcgen 下载参考 https://github.com/thkukuk/rpcsvc-proto/releases

3.5.4 源码安装

实际上，MySQL 源码包就是一大堆源代码程序，是由程序员按照特定的格式和语法编写出来的。由于源码包的安装需要把源代码编译为二进制代码，因此安装时间较长。

1. 解压软件包

#创建软件目录
[root@jeames ~]# mkdir -p /soft

2. 上传软件包

#通过 Xftp 工具或者 SecureCRT 自带的 Sftp 上传软件包
mysql-boost-8.0.27.tar.gz

3. 校验安装包

[root@jeames ~]# cd /soft
[root@jeames soft]# md5sum mysql-boost-8.0.27.tar.gz
80310c5a1b24145fa072927ab99a4c0d mysql-boost-8.0.27.tar.gz
注意:md5sum 是 linux 下的 shell 命令,其功能是计算检验 MD5 校验码,是为了 MySQL 安装包是否损坏。

4. 解压缩安装包

[root@jeames soft]# tar -zxf mysql-boost-8.0.27.tar.gz

5. 编译安装

(1)初始化
[root@jeames soft]# cd mysql-8.0.27
cmake -DCMAKE_INSTALL_PREFIX=/usr/local/mysql \
-DMYSQL_DATADIR=/usr/local/mysql/data \
-DSYSCONFDIR=/etc \
-DWITH_INNOBASE_STORAGE_ENGINE=1 \
-DWITH_PARTITION_STORAGE_ENGINE=1 \
-DWITH_FEDERATED_STORAGE_ENGINE=1 \
-DWITH_BLACKHOLE_STORAGE_ENGINE=1 \
-DWITH_MYISAM_STORAGE_ENGINE=1 \
-DENABLED_LOCAL_INFILE=1 \
-DENABLE_DTRACE=0 \
-DDEFAULT_CHARSET=utf8mb4 \

-DDEFAULT_COLLATION= utf8mb4_general_ci \
-DWITH_EMBEDDED_SERVER= 1 \
-DDOWNLOAD_BOOST= 1 \
-DFORCE_INSOURCE_BUILD= 1 \
-DWITHOUT_PARTITION_STORAGE_ENGINE= 0 \
-DCMAKE_C_COMPILER= /usr/bin/gcc \
-DCMAKE_CXX_COMPILER= /usr/bin/g++ \
-DWITH_BOOST= /soft/mysql-8.0.27/boost

（2）编译

[root@jeames mysql-8.0.27]# make -j4

（3）安装

[root@jeames mysql-8.0.27]# make install

图 3-26 所示为初始化成功的界面，通过编译和安装最终安装完成。

图 3-26　源码编译成功后的界面

3.5.5　安装 MySQL

源码编译后与二进制包部署一样，只需要配置好参数文件，初始化后即可安装 MySQL。

1. 用户及组

groupadd mysql
useradd -g mysql mysql
chown -R mysql:mysql /usr/local/mysql

2. 参数文件

cat > /etc/my.cnf << "EOF"
[mysqld]

```
basedir= /usr/local/mysql
datadir= /usr/local/mysql/data
port= 3306
server_id= 80273306
log-bin
skip-name-resolve
character_set_server= utf8mb4
# default-time-zone = '+ 8:00'
log_timestamps = SYSTEM
EOF
```

3. MySQL 初始化

```
/usr/local/mysql/bin/mysqld --initialize-insecure
--basedir= /usr/local/mysql --datadir= /usr/local/mysql/data --user= mysql
```

4. 环境变量设置

```
echo "export PATH= $ PATH:/usr/local/mysql/bin" >> /root/.bashrc
source /root/.bashrc
```

3.5.6 启动 MySQL

启动 MySQL mysqld_safe &。
登录 MySQL,默认密码为空 mysql-uroot-p。
关闭 MySQL mysqladmin-uroot-p shutdown。

3.6 MySQL 版本升级

MySQL 升级过程大致分为两部分,升级数据字典和升级服务。升级方式分为两种:原地升级和逻辑升级。这两种升级方式,本质没有什么区别的,只是在对数据文件的处理上,有些区别而已。

原地升级是直接复制数据文件,而逻辑升级对数据文件的处理方式是通过逻辑导出、导入,需要用到 mysqldump。逻辑升级这种方式在数据量比较大的情况下花费时间比较长,生产环境建议原地升级,这里演示了 Rhel8 系统下 5.7 版本升级到 8.0 的全部过程。升级数据库版本可能会有各种原因,但升级的前提一定要做到未雨绸缪,尽量减少对业务的影响。

3.6.1 为什么升级

(1)基于安全考虑,可以避免低版本没有彻底解决的 BUG。
(2)基于性能和稳定性考虑:MGR 复制、分区表等功能,性能提升大。
(3)原始环境中版本太多,统一版本。
(4)8.0 版本基本已到稳定期,新特性多,如窗口函数、支持 json 等。

3.6.2 升级注意事项

(1)支持从 MySQL5.7 升级到 8.0,注意仅支持 GA 版本之间的升级。
(2)不支持跨大版本的升级,如从 5.6 升级到 8.0 是不支持的。
(3)建议升级大版本前先升级到当前版本的最近小版本,如 5.6 先升级到 5.7 后再升级 8.0。
(4)注意字符集设置、密码认证插件变更和 sql_mode 设置。
(5)高可用架构下需要先升级从库,再逐步升级主库。
(6)是否需要手动升级系统表。MySQL8.0.16 版本之前,需要手动执行 mysql_upgrade 来完成该步骤的升级,在 MySQL8.0.16 版本及之后由 mysqld 来完成该步骤的升级。

3.6.3 环境准备

本次数据库升级在同一台服务器进行,是基于 RHEL8 操作系统,升级前做好备份,便于升级失败回退。

1. 操作系统

[root@jeames ~]# cat /etc/redhat‐release
Red Hat Enterprise Linux release 8.1 (Ootpa)

2. 关闭防火墙

查看防火墙状态
[root@jeames ~]# systemctl status firewalld
关闭防火墙
[root@jeames ~]# systemctl stop firewalld
取消开机自启动
[root@jeames ~]# systemctl disable firewalld
Removed /etc/systemd/system/multi‐user.target.wants/firewalld.service.
Removed /etc/systemd/system/dbus‐org.fedoraproject.FirewallD1.service.

3. selinux 关闭

修改参数文件/etc/sysconfig/selinux 中 SELINUX 的值为 disabled

[root@jeames ~]# sed -i 's/SELINUX\=enforcing/SELINUX\=disabled/g' /etc/selinux/config

修改完成后需要重启服务器才生效。

3.6.4 yum 安装依赖

本地 yum 配置前面均有介绍，此处不再累赘，如果是测试环境可以通过连接外网，配置在线 yum。

[root@jeames ~]# yum install libaio
[root@jeames ~]# yum -y install perl perl-devel
[root@jeames ~]# yum install libncurses*
[root@jeames ~]# yum -y install autoconf
[root@jeames ~]# yum -y install numactl.x86_64

3.6.5 安装 MySQL5.7

通过二进制的方法先部署 MySQL5.7，注意环境变量的配置。

1. 二进制包解压

[root@jeames ~]# mkdir -p /usr/local/mysqlsoft
[root@jeames ~]# cd /soft/
tar -zxvf mysql-5.7.30-linux-glibc2.12-x86_64.tar.gz -C /usr/local/mysqlsoft

2. 快捷方式创建

ln -s /usr/local/mysqlsoft/mysql-5.7.30-linux-glibc2.12-x86_64 /usr/local/mysql57/mysql5730

3. 创建用户组

[root@jeames ~]# groupadd mysql
[root@jeames ~]# useradd -r -g mysql mysql
[root@jeames ~]# chown -R mysql:mysql /usr/local/mysqlsoft

4. 初始化(5.7 版本)

/usr/local/mysql57/mysql5730/bin/mysqld --initialize-insecure --user=mysql \
    --basedir=/usr/local/mysql57/mysql5730
    --datadir=/usr/local/mysql57/mysql5730/data

5. 配置 5.7 版本的环境变量

[root@jeames ~]# echo "export PATH=$PATH:/usr/local/mysql57/mysql5730/bin" >> /root/.bashrc

[root@jeames ~]# source /root/.bashrc

6. 启动数据库

[root@jeames ~]# mysqld_safe &

7. 新增远程用户及修改密码

mysql> select user,host,authentication_string from mysql.user

mysql> grant all on *.* to root@'%' identified by 'root' with grant option

mysql> flush privileges

修改本地密码

mysql> set password for root@'localhost'=password('root')

mysql> flush privileges

3.6.6 数据库升级

1. 升级前注意事项

(1) 关闭数据库之前，首先使用 show processlist 检查链接是否已经关闭。

(2) 备份源库的数据（包括当前的数据库和日志文件），建议使用物理备份，甚至如果条件允许，直接使用冷备模式。

(3) 源库升级检查：mysqlcheck -u root -p --all-database --check-upgrade。

(4) 设置了 innodb_fast_shutdown 参数为 0，innodb_fast_shutdown 参数用于设置 MySQL InnoDB 引擎的关闭模式，设置为 0 时，InnoDB 关闭慢，需要清除所有的 undo log（除了 XA prepare 的事务），完成 change buffer 的合并，将脏页刷盘，关闭 redo log。

(5) sql_mode 支持问题，8.0 版本 sql_mode 不支持 NO_AUTO_CREATE_USER，要避免配置的 sql_mode 中带有 NO_AUTO_CREATE_USER。

(6) 注意字符集设置，为了避免新旧对象字符集不一致的情况，在配置文件时，可将字符集和校验规则设置为旧版本的字符集和比较规则。

(7) 密码认证插件变更，为了避免链接问题，可仍采用 5.7 的 mysql_native_password 认证插件。

(8) 关键字在 MySQL 8.0 版本中新增了一些关键字，而我们都知道关键字是无法使用的，如果在之前版本中使用了，那么就会报错。

2. 安装 MySQL8

通过二进制的方法先部署 MySQL8.0 的版本。

(1) 解压安装包

tar -Jxf mysql-8.0.19-linux-glibc2.12-x86_64.tar.xz -C /usr/local/mysqlsoft

(2) 快捷方式创建

[root@jeames ~]# mkdir -p /usr/local/mysql80

ln -s /usr/local/mysqlsoft/mysql-8.0.19-linux-glibc2.12-x86_64 /usr/local/mysql80/mysql8019

(3) MySQL8 初始化

[root@jeames ~]# chown -R mysql.mysql /usr/local/mysql80

[root@jeames ~]# /usr/local/mysql80/mysql8019/bin/mysqld --initialize-insecure --user=mysql \

--basedir=/usr/local/mysql80/mysql8019

--datadir=/usr/local/mysql80/mysql8019/data

3. 升级关键步骤

(1) 查找参数文件

mysql --help | grep 'my.cnf'

/etc/my.cnf /etc/mysql/my.cnf /usr/local/mysql/etc/my.cnf ~/.my.cnf

(2) 修改参数文件

[root@jeames ~]# vi /etc/my.cnf

[mysqld]

根目录为 8.0 的根目录

basedir=/usr/local/mysql80/mysql8019

数据文件目录为 5.7 的目录

datadir=/usr/local/mysql57/mysql5730/data

端口保持 5.7 的端口

port=3306

错误日志保持 5.7 版本的

log-error=/usr/local/mysql57/mysql5730/data/log.err

8.0 的 server_id

server_id=80193306

log-bin

密码验证方式

default_authentication_plugin=mysql_native_password

character_set_server=utf8mb4

(3)配置 MySQL8.0 的环境变量

echo " export PATH = $ PATH:/usr/local/mysql80/mysql8019/bin " >> /root/.bashrc

source /root/.bashrc

此处,记得要删除原来的 5.7 版本环境变量,配置 8.0 版本变量,重启下服务器,后再启动 MySQL8。

(4)升级后 MySQL8.0 重置密码

mysql> alter user root@'localhost' identified with mysql_native_password by 'root'

mysql> create user root@'%' identified with mysql_native_password by 'root'

mysql> grant all on *.* to root@'%' with grant option

mysql> flush privileges

其实升级的关键就是使用根目录 MySQL8.0 的路径,其他的使用升级前 5.7 的所有路径,升级成功后如图 3-27 所示。

```
mysql> status
--------------
mysql  Ver 8.0.19 for linux-glibc2.12 on x86_64 (MySQL Community Server - GPL)

Connection id:          15
Current database:
Current user:           root@localhost
SSL:                    Not in use
Current pager:          stdout
Using outfile:          ''
Using delimiter:        ;
Server version:         8.0.19 MySQL Community Server - GPL
Protocol version:       10
Connection:             Localhost via UNIX socket
Server characterset:    utf8mb4
Db     characterset:    utf8mb4
Client characterset:    utf8mb4
Conn.  characterset:    utf8mb4
UNIX socket:            /tmp/mysql.sock
Binary data as:         Hexadecimal
Uptime:                 1 min 57 sec
```

图 3-27 MySQL 5.7 版本升级到 8.0 版本

3.7 MySQL 客户端工具

MySQL 的管理维护工具非常多,除了系统自带的命令行管理工具之外,还有许多其他的图形化管理工具。工欲善其意,必先利其器,好用的工具能成倍提高工作效率,这里介绍几个经常使用的 MySQL 图形化管理工具。

3.7.1 MySQL workbench

MySQL Workbench 是一个统一的可视化开发和管理平台,该平台提供了许多高级工具,可支持数据库建模和设计、查询开发和测试、服务器配置和监视、用户和安全管理、备份和恢复自动化、审计数据检查以及向导驱动的数据库迁移。

MySQL Workbench 是 MySQL AB 发布的可视化的数据库设计软件,支持 Windows、Linux、Mac 主流的操作系统,使用起来非常好。MySQL Workbench 为数据库管理员、程序开发者和系统规划师提供可视化设计、模型建立以及数据库管理功能,其界面如图 3-28 所示。

官网下载地址:https://dev.mysql.com/downloads/workbench/。

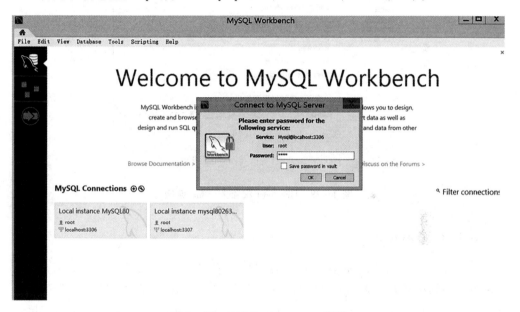

图 3-28　MySQL Workbench 界面

3.7.2 DBeaver

DBeaver 是一个通用的 MySQL 数据库管理工具,支持 MySQL、PostgreSQL、Oracle、DB2、MSSQL、Sybase、Mimer、HSQLDB、Derby 以及其他兼容 JDBC 的数据库,是基于 Java 进行开发的。DBeaver 提供一个图形界面用来查看数据库结构、执行 SQL 查询和脚本,浏览和导出数据,处理 BLOB/CLOB 数据,修改数据库结构等。

DBeaver 是一个跨平台的数据库管理工具,支持 Windows、Linux 和 macOS。它有两个版本,企业版和社区版,对于个人开发者来说,社区版的功能已经足够用。

可以通过 DBeaver 官方下载安装包,也可以通过 GitHub 下载 release 版本,界面如图 3-29 所示。

因为 DBeaver 是基于 Maven 构建的,数据库驱动也就是链接数据库的 JDBC 驱动,是通过 Maven 仓库下载的,配置 Maven 镜像后续,后续下载数据库驱动包即可使用。

选择"首选项"→"Maven",添加阿里云镜像地址,配置完成后,记得把阿里云镜像仓库置顶。

图 3-29 DBeaver 配置 MySQL 连接

3.7.3 Navicat

Navicat 是一套可创建多个连接的数据库管理工具,用以方便管理 MySQL、Oracle、PostgreSQL、SQLite、SQL Server、MariaDB 和 MongoDB 等不同类型的数据库,它与阿里云、腾讯云、华为云、Amazon RDS、Amazon Aurora、Amazon Redshift、Microsoft Azure、Oracle Cloud 和 MongoDB Atlas 等云数据库兼容。大家可以创建、管理和维护数据库。

Navicat 的功能足以满足专业开发人员的所有需求,但是对数据库服务器初学者来说又简单易操作。Navicat 的用户界面(GUI)设计良好,可以安全且简单的方法创建、组织、访问和共享信息,如图 3-30 所示。

3.7.4 phpMyAdmin

phpMyAdmin 是用 PHP 脚本写的 MySQL 数据库管理软件,是使用 Web 图形模式直接管理 MySQL 数据库的工具。phpMyAdmin 可以用来创建、修改、删除数据库和数据表;也可以用来创建、修改、删除数据记录;可以用来导入和导出整个数据库;还可以完成许多其他的 MySQL 系统管理任务。

官网地址：https://www.phpmyadmin.net/。

phpMyAdmin 安装包下载：https://www.phpmyadmin.net/downloads/，界面如图 3-31 所示。

图 3-30　使用 Navicat 配置 MySQL 查询

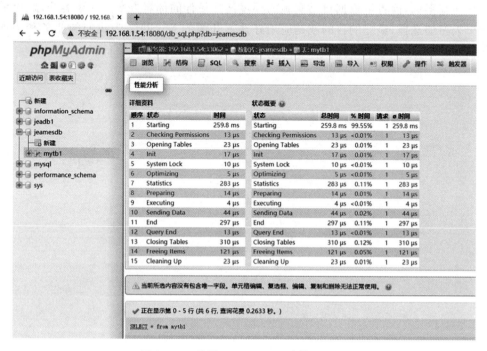

图 3-31　使用 phpMyAdmin 连接 MySQL

3.7.5 SQLyog

SQLyog是一个易于使用的、快速而简洁的图形化管理MYSQL数据库的工具，它能够在任何地点有效地管理您的数据库。

SQLyog是业界著名的Webyog公司出品的一款简洁高效、功能强大的图形化MySQL数据库管理工具。使用SQLyog可以快速直观地让您从世界的任何角落通过网络来维护远端的MySQL数据库。

SQLyog完美支持64位WIN7及以后版本的系统，可以连接到指定的MYSQL主机，支持使用HTTP管道以及/SSH/SSL，可创建新的表、视图、存储过程、函数、触发器及事件，支持删除及截位数据库。支持转储数据库，将数据库保存到SQL，编辑功能可以查找/替换指定内容，可列出全部或匹配标记，管理由SQLyog创建的任务，使用各自的任务向导创建任务等，解除了使用限制，安全免费！

小 结

对于社区版，任何人都能从Internet下载MySQL软件，而无需支付任费用，并且"开放源码"意味着任何人都可以使用和修改该软件。本章基于Linux及Windows环境讲解了全部二进制、源码、图形化等方式的部署过程。大多数案例均来自客户实施案例，读者可根据自己实施的环境选择最佳的部署方案。对于客户端，可依据个人习惯选择，最终的目的就是应用好这款开源的MySQL，为我们的业务和生产服务。

Part 2　运维篇

第4章　DBMS基本管理

4.1　数据类型

MySQL数据类型定义了列中可以存储什么数据以及该数据怎样存储的规则，数据库中的每个列都应该有适当的数据类型，用于限制或允许该列中存储的数据，如果使用错误的数据类型可能会严重影响应用程序的功能和性能。所以在设计表时，应该特别重视数据列所用的数据类型。

MySQ支持多种类型，大致可以分为三类：数值、日期/时间和字符串(字符)类型。

4.1.1　数值类型

MySQL支持所有标准SQL数值数据类型，整数类型包括TINYINT、SMALLINT、MEDIUMINT、INT、BIGINT，浮点数类型包括FLOAT和DOUBLE，定点数类型为DECIMAL。

如果使用的数据类型超出了数值类型的范围，则MySQL会抛出相应的错误。因此在实际使用的时候，应该首先确认好数据的取值范围，然后根据确认的结果选择合适的整数类型。

1. 整数类型

整数类型的存储和范围见表4-1所列。

表4-1　整数类型的存储和范围

类型	大小	范围(有符号)	范围(无符号)	用途
TINYINT	1 Bytes	(-128,127)	(0,255)	小整数值
SMALLINT	2 Bytes	(-32 768,32 767)	(0,65 535)	大整数值
MEDIUMINT	3 Bytes	(-8 388 608,8 388 607)	(0,16 777 215)	大整数值
INT 或 INTEGER	4 Bytes	(-2 147 483 648, 2 147 483 647)	(0,4 294 967 295)	大整数值
BIGINT	8 Bytes	(-9,223,372,036,854,775,808, 9 223 372 036 854 775 807)	(0,18 446 744 073 709 551 615)	极大整数值

2. 带小数类型

FLOAT(单精度)、DOUBLE(双精度)、DECIMAL(定点数)

举例:DECIMAL(P,D)。
P 是表示有效数字数的精度,P 范围为 1~65。
D 是表示小数点后的位数,D 的范围是 0~30。
MySQL 要求 D 小于或等于(<=)P。
DECIMAL(4,2)如:23.46。

4.1.2 日期/时间类型

表示时间值的日期和时间类型为 DATETIME、DATE、TIMESTAMP、TIME 和 YEAR,每种日期和时间类型都有一个有效值范围,如果超出这个有效值范围,则会以 0 进行存储,见表 4-2 所列。

表 4-2 日期/时间类型的存储和范围

类型	大小	范围	格式	用途
DATE	3 Bytes	1000-01-01/9999-12-31	YYYY-MM-DD	日期值
TIME	3 Bytes	'-838:59:59'/'838:59:59'	HH:MM:SS	时间值或持续时间
YEAR	1 Bytes	1901/2155	YYYY	年份值
DATETIME	8 Bytes	'1000-01-01 00:00:00' 到 '9999-12-31 23:59:59'	YYYY-MM-DD hh:mm:ss	混合日期和时间值
TIMESTAMP	4 Bytes	'1970-01-01 00:00:01' UTC 到 '2038-01-19 03:14:07' UTC	YYYY-MM-DD hh:mm:ss	混合日期和时间值,时间戳

(1)TIME 类型:用于只需要时间信息的值,在存储时需要 3 个字节。格式为 HH:MM:SS。HH 表示小时,MM 表示分钟,SS 表示秒。

(2)DATE 类型:DATE 类型用于仅需要日期值时,没有时间部分,在存储时需要 3 个字节,日期格式为'YYYY-MM-DD',其中 YYYY 表示年,MM 表示月,DD 表示日。

MySQL 允许"不严格"语法:任何标点符号都可以用作日期部分之间的间隔符。例如,'93-11-31'、'93.11.31'、'93/11/31'和'98@11@31'是等价的,这些值也可以正确地插入数据库。

(3)TIMESTAMP 与 DATETIME 除了存储字节和支持的范围不同外,还有一个最大的区别是:

DATETIME 在存储日期数据时,按实际输入的格式存储,即输入什么就存储什么,与时区无关;而 TIMESTAMP 值的存储是以 UTC(世界标准时间)格式保存的,存储时对当前时区进行转换,检索时再转换回当前时区。即查询时,根据当前时区的不同,显示的时间值是不同的。

4.1.3 字符串类型

在 MySQL 中,字符串类型可以存储文本字符串数据,也可以存储一些图片、音频和视频数据,也就是二进制数据。因此在 MySQL 中,字符串类型可以分为文本字符串类型和二进制字符串类型。

1. 文本字符串类型

MySQL 中,文本字符串总体上分为 CHAR、VARCHAR、TINYTEXT、TEXT、MEDIUMTEXT、LONGTEXT、ENUM、SET 和 JSON 等类型,每种存储类型所占的存储空间见表 4-3 所列。

表 4-3 文本字符串类型的存储和范围

类型	大小	用途
CHAR	0-255 bytes	定长字符串
VARCHAR	0-65 535 bytes	变长字符串
TINYTEXT	0-255 bytes	短文本字符串
TEXT	0-65 535 bytes	长文本数据
MEDIUMTEXT	0-16 777 215 bytes	中等长度文本数据
LONGTEXT	0-4 294 967 295 bytes	极大文本数据
ENUM	1 或 2 个字节,取决于枚举值的数目(最大值为 65 535)	枚举类型,只能有一个枚举字符串值
SET	1、2、3、4 或 8 个字节,取决于集合成员的数量(最多 64 个成员)	字符串对象可以有零个或多个 SET 成员

(1)char(n)和 varchar(n)中括号中 n 代表字符的个数,并不代表字节个数,比如 CHAR(30)就可以存储 30 个字符。

(2)CHAR 和 VARCHAR 类型类似,但它们保存和检索的方式不同。它们的最大长度和是否尾部空格被保留等方面也不同。在存储或检索过程中不进行大小写转换。

2. 二进制字符串类型

MySQL 中的二进制字符串类型主要存储一些二进制数据,比如可以存储图片、音频和视频等二进制数据,MySQL 中的二进制字符串有 BIT、BINARY、VARBINARY、TINYBLOB、BLOB、MEDIUMBLOB 和 LONGBLOB。

表 4-4 中列出了 MySQL 中的二进制数据类型,括号中的 M 表示可以为其指定长度。

表 4-4 二进制字符串类型

类型名称	说明	存储需求
BIT(M)	位字段类型	大约(M+7)/8 字节
BINARY(M)	固定长度二进制字符串	M 字节
VARBINARY(M)	可变长度二进制字符串	M+1 字节
TINYBLOB(L)	非常小的 BLOB	L+1 字节,在此 L<2^8
BLOB(L)	小 BLOB	L+2 字节,在此 L<2^16
MEDIUMBLOB(L)	中等大小的 BLOB	L+3 字节,在此 L<2^24
LONGBLOB(L)	非常大的 BLOB	L+4 字节,在此 L<2^32

温馨提示:

(1)BIT 类型中存储的是二进制值

BIT 数据类型用来保存位字段值,例如以二进制的形式保存数据 13,13 的二进制形式为 1101,在这里需要位数至少为 4 位的 BIT 类型,即可以定义列类型为 BIT(4),大于二进制 1111 的数据是不能插入 BIT(4)类型的字段中的。

(2)BINARY 和 VARBINARY 类似于 CHAR 和 VARCHAR,不同的是它们包含二进制字符串而不是非二进制字符串。也就是说,它们包含字节字符串而不是字符字符串。这说明它们没有字符集,并且排序和比较是基于列值字节的数值值。

(3)BLOB 是一个二进制大对象,其可以容纳可变数量的数据。有 4 种 BLOB 类型:TINYBLOB、BLOB、MEDIUMBLOB 和 LONGBLOB,它们区别在于可容纳存储范围不同。

(4)BLOB 和 TEXT 的区别

BLOB 列存储的是二进制字符串(字节字符串),TEXT 列存储的是非进制字符串(字符字符串)。BLOB 列是字符集,并且排序和比较基于列值字节的数值;TEXT 列有一个字符集,并且根据字符集对值进行排序和比较。

(5)需要注意的是,在实际运维工作中,往往不会在 MySQL 数据库中使用 BLOB 类型存储大对象数据,通常会将图片、音频和视频文件存储到服务器的磁盘上,并将图片、音频和视频的访问路径存储到 MySQL 中。

4.2 数据库管理

MySQL 是一个多用户、多实例数据库,即单进程多线程架构。我们可以在登录 MySQL 服务后,使用 create 命令创建数据库,语法是:CREATE DATABASE 数据库名。

注意：db1、db2、db3 分别为数据库名。

♯创建数据库

mysql> create database db1

mysql> create database db2 character set utf8mb4

mysql> create database db3 charset utf8mb4

♯删除数据库

-- windows 系统

mysql> drop database db1

-- Linux 系统

mysql> mysqladmin - uroot - proot - h192.168.1.5 - P3306 drop db2

mysql> mysql - uroot - proot - h192.168.1.5 - P3306 - e "drop database db2"

♯选择数据库

mysql> use db1

Database changed

执行以上命令后，就已经成功选择了 db1 数据库，在后续的操作中都会在 db1 数据库中执行。

♯查看创建数据库语句

mysql> show create databasedb1

(1)MySQL8.0 之前默认的数据库字符集是 latin1，从 8.0 开始，默认就是 utf8mb4 字符集。

(2)utf8mb4 支持 BMP 和补充字符，MySQL8 中建议使用 utf8mb4。

(3)创建的数据库字符集与参数设置有关。

(4)在删除数据库过程中，务必要十分谨慎，因为在执行删除命令后，所有数据将会消失。

4.3 表的管理

数据表是数据库的重要组成部分，每一个数据库都是由若干个数据表组成的。换句话说，没有数据表就无法在数据库中存放数据。本节将详细介绍数据表的基本操作，主要包括创建数据表、查看数据表结构、修改数据表和删除数据表等。

4.3.1 创建数据表

在 MySQL 中，可以使用 CREATE TABLE 语句创建表。CREATE TABLE 命令语法比较多，其主要是由表创建定义(create - definition)、表选项(table - options)和分区选项(partition - options)所组成的。

其语法格式为：

CREATE TABLE <表名>（[表定义选项]）[表选项][分区选项]；

其中，[表定义选项]的格式为：

<列名 1> <类型 1>[,…]<列名 n> <类型 n>

使用 CREATE TABLE 创建表时，必须指定以下信息：

(1)要创建的表的名称不区分大小写(参数设定了大小写不敏感)，不能使用 SQL 语言中的关键字，如 DROP、ALTER、INSERT 等。

(2)数据表中每个列(字段)的名称和数据类型，如果创建多个列，要用逗号隔开。

(3)指定操作在哪个数据库中进行，如果没有选择数据库，就会抛出 No database selected 的错误。

(4)用于创建给定名称的表，必须拥有表 CREATE TABLE 的权限。

创建学生表 tb_student，结构见表 4-5 所列。

表 4-5 学生表 tb_student

字段名称	数据类型	允许 NULL 值	约束	备注
studentNo	CHAR(10)	非空	主键	学号
studentName	VARCHAR(10)	非空		姓名
sex	CHAR(2)			性别
birthday	DATE			出生日期
native	VARCHAR(20)			籍贯
nation	VARCHAR(10)			民族,默认值:汉
classNo	CHAR(6)			所属班级

选择创建表的数据库 db_school，创建 tb_student 数据表，输入的 SQL 语句如下所示：

mysql> USE test_db

create table if not exists tb_student (
studentNo CHAR(10) not NULL primary key comment '学号'
studentName VARCHAR(10) NOT null comment '姓名'
sex CHAR(2) comment '性别'
birthday date comment '出生日期'
native VARCHAR(20) comment '籍贯'
nation VARCHAR(10) DEFAULT '汉' comment '民族'

classNo CHAR(6) comment '所属班级'
) ENGINE= InnoDB comment '学生表'

使用 DESC 查看表 tb_student 的表结构,SQL 语句和运行结果如下:

mysql> desc tb_student

Field	Type	Null	Key	Default	Extra
studentNo	char(10)	NO	PRI	NULL	
studentName	varchar(10)	NO		NULL	
sex	char(2)	YES		NULL	
birthday	date	YES		NULL	
native	varchar(20)	YES		NULL	
nation	varchar(10)	YES		汉	
classNo	char(6)	YES		NULL	

其中,各个字段的含义如下:

Null:表示该列是否可以存储 NULL 值。

Key:表示该列是否已编制索引。PRI 表示该列是表主键的一部分,UNI 表示该列是 UNIQUE 索引的一部分,MUL 表示在列中某个给定值允许出现多次。

Default:表示该列是否有默认值,如果有,值是多少。

Extra:表示可以获取的与给定列有关的附加信息,如 AUTO_INCREMENT 等。

同 DESC 相比,SHOW CREATE TABLE 展示的内容更加丰富,它可以查看表的存储引擎和字符编码,SHOW CREATE TABLE 的语法格式如下:

SHOW CREATE TABLE <表名>

在 SHOW CREATE TABLE 语句的结尾处(分号前面)添加\g 或者\G 参数可以改变展示形式。

mysql> SHOW CREATE TABLE tb_student\G
*************************** 1. row ***************************
       Table: tb_student
Create Table: CREATE TABLE 'tb_student' (
  'studentNo' char(10) NOT NULL COMMENT '学号'
  'studentName' varchar(10) NOT NULL COMMENT '姓名'
  'sex' char(2) DEFAULT NULL COMMENT '性别'
  'birthday' date DEFAULT NULL COMMENT '出生日期'
  'native' varchar(20) DEFAULT NULL COMMENT '籍贯'

'nation' varchar(10) DEFAULT '汉' COMMENT '民族'

'classNo' char(6) DEFAULT NULL COMMENT '所属班级'

PRIMARY KEY ('studentNo')

) ENGINE = InnoDB DEFAULT CHARSET = utf8mb4 COLLATE = utf8mb4_0900_ai_ci COMMENT = '学生表'

1 row in set (0.00 sec)

大家可使用 CREATE TALBE SELECT 语句将查询结果转存到一个新表中,且不会自动创建任何索引及约束。

mysql> create table tb_student2 select * from tb_student where 1= 3

4.3.2 修改数据表

修改数据表的前提是数据库中已经存在该表。修改表指的是修改数据库中已经存在的数据表的结构。修改数据表的操作也是数据库管理中必不可少的,就像画素描一样,画多了可以用橡皮擦掉,画少了可以用笔加上。在 MySQL 中,可以使用 ALTER TABLE 语句来改变原有表的结构,例如增加或删减列、更改原有列类型、重新命名列或表等。

其语法格式如下:

ALTER TABLE ＜表名＞［修改选项］

修改选项的语法格式如下:

｛ADD COLUMN ＜列名＞ ＜类型＞

｜CHANGE COLUMN ＜旧列名＞ ＜新列名＞ ＜新列类型＞

｜ALTER COLUMN ＜列名＞｛SET DEFAULT ＜默认值＞｜DROP DEFAULT｝

｜MODIFY COLUMN ＜列名＞ ＜类型＞

｜DROP COLUMN ＜列名＞

｜RENAME TO ＜新表名＞

｜CHARACTER SET ＜字符集名＞

｜COLLATE ＜校对规则名＞｝

1. 修改表名

MySQL 通过 ALTER TABLE 语句来实现表名的修改,语法规则如下:

ALTER TABLE ＜表名＞［修改选项］

其中,TO 为可选参数,使用与否均不影响结果。

案例:使用 ALTER TABLE 将数据表 tb_student 改名为 backup_tb_student

mysql> alter table tb_student2 rename to backup_tb_student

修改表名并不修改表的结构,因此修改名称后的表和修改名称前的表的结构是相同的,用户可以使用 DESC 命令查看修改后的表结构。

2. 修改表字符集

MySQL 通过 ALTER TABLE 语句来实现表字符集的修改,语法规则如下:

ALTER TABLE 表名 [DEFAULT] CHARACTER SET <字符集名> [DEFAULT] COLLATE <校对规则名>

其中,DEFAULT 为可选参数,使用与否均不影响结果

案例:将数据表 tb_student 的字符集修改为 gb2312,校对规则修改为 gb2312_chinese_ci

mysql> ALTER TABLE tb_student CHARACTER SET gb2312 DEFAULT COLLATE gb2312_chinese_ci

3. 数据表添加字段

MySQL 数据表是由行和列构成的,通常把表的"列"称为字段(Field),把表的"行"称为记录(Record)。随着业务的变化,可能需要在已有的表中添加新的字段。MySQL 允许在开头、中间和结尾处添加字段。

(1)在末尾添加字段

一个完整的字段包括字段名、数据类型和约束条件。MySQL 添加字段的语法格式如下:

ALTER TABLE <表名> ADD <新字段名><数据类型>[约束条件]

语法格式:

<表名> 为数据表的名字

<新字段名> 为所要添加的字段的名字

<数据类型> 为所要添加的字段能存储数据的数据类型

[约束条件] 是可选的,用来对添加的字段进行约束

这种语法格式默认在表的最后位置(最后一列的后面)添加新字段。

案例:给数据表 tb_student 添加一个 varchar(16)类型的字段 department,要求设置其默认值为"城市学院"

mysql> ALTER TABLE tb_student ADD department varchar(16) DEFAULT '城市学院'

(2)在开头添加字段

MySQL 默认在表的最后位置添加新字段,如果希望在开头位置(第一列的前面)添加新字段,那么可以使用 FIRST 关键字,语法格式如下:

ALTER TABLE <表名> ADD <新字段名><数据类型>[约束条件] FIRST

FIRST 关键字一般放在语句的末尾

案例：在表 tb_student 的第一列添加 INT 类型的字段 id，要求其不能为 NULL

mysql> ALTER TABLE tb_student ADD id INT FIRST

（3）在中间位置添加字段

MySQL 除了允许在表的开头位置和末尾位置添加字段外，还允许在中间位置（指定的字段之后）添加字段，此时需要使用 AFTER 关键字，语法格式如下：

ALTER TABLE ＜表名＞ ADD ＜新字段名＞ ＜数据类型＞［约束条件］AFTER ＜已经存在的字段名＞

AFTER 的作用是将新字段添加到某个已有字段后面。

注意，只能在某个已有字段的后面添加新字段，不能在它的前面添加新字段。

案例：在 tb_student 表中添加 INT 类型的字段 age，age 字段位于 studentName 字段的后面

mysql> ALTER TABLE tb_student ADD age INT(4) AFTER studentName

4. 数据表修改字段

（1）修改字段名称

MySQL 中修改表字段名的语法规则如下：

ALTER TABLE ＜表名＞ CHANGE ＜旧字段名＞ ＜新字段名＞ ＜新数据类型＞

语法格式说明：

旧字段名：指修改前的字段名。

新字段名：指修改后的字段名。

新数据类型：指修改后的数据类型，如果不需要修改字段的数据类型，可以将新数据类型设置成与原来一样，但数据类型不能为空。

CHANGE 也可以只修改数据类型，实现和 MODIFY 同样的效果，方法是将 SQL 语句中的"新字段名"和"旧字段名"设置为相同的名称，只改变"数据类型"。

由于不同类型的数据在机器中的存储方式及长度并不相同，修改数据类型可能会影响数据表中已有的数据记录，因此，当数据表中已经有数据时，不要轻易修改数据类型。

案例：修改表 tb_student 的结构，将 studentName 字段名称改为 Name，同时将数据类型变为 varchar(20)。

mysql> ALTER TABLE tb_student CHANGE studentName Name varchar(20)

（2）修改字段数据类型

修改字段的数据类型就是把字段的数据类型转换成另一种数据类型。

在 MySQL 中修改字段数据类型的语法规则如下：

ALTER TABLE ＜表名＞ MODIFY ＜字段名＞ ＜数据类型＞

语法格式：
表名：指要修改数据类型的字段所在表的名称
字段名：指需要修改的字段
数据类型：指修改后字段的新数据类型

案例：修改 tb_student 中的字段 age，并将其数据类型更改为 TINYINT
mysql> ALTER TABLE tb_student MODIFY age TINYINT

(3) 其他修改

案例 1：将 tb_student 表中的字段 department 的默认值改为'环化学院'
mysql> ALTER TABLE tb_student ALTER department SET DEFAULT '环化学院'

案例 2：将 tb_student 表中的字段 department 的默认值删除
mysql> ALTER TABLE tb_student ALTER department DROP DEFAULT

案例 3：将 tb_student 表中的字段 department 的数据类型更改为 varchar(20)，取值不允许为空，并将此字段移至字段 birthday 之后
mysql> ALTER TABLE tb_student MODIFY department varchar(20) not null after birthday

5. 数据表删除字段

删除字段是将数据表中的某个字段从表中移除，语法格式如下：
ALTER TABLE <表名> DROP <字段名>
其中，"字段名"指需要从表中删除的字段的名称。
如果删除的主键是自增时，先删除自增长，增删除主键，再按以下方式处理：
mysql> ALTER TABLE tb_student2 MODIFYid int not null; //删除自增长
mysql> Alter table tb_student2 drop primary key;//删除主建
mysql> ALTER TABLE tb_student2 DROP id;

案例：删除数据表 tb_student 中的字段 id
mysql> ALTER TABLE tb_student DROP id

4.3.3 删除数据表

在 MySQL 数据库中，对于不再需要的数据表，我们可以将其从数据库中删除。在删除表的同时，表的结构和表中所有的数据都会被删除，因此在删除数据表之前最好先备份，以免造成无法挽回的损失。

使用 DROP TABLE 语句可以删除一个或多个数据表，语法格式如下：
DROP TABLE [IF EXISTS] 表名1 [,表名2，表名3...]
用户必须拥有执行 DROP TABLE 命令的权限，否则数据表不会被删除。
表被删除时，用户在该表上的权限不会自动删除。

IF EXISTS 用于在删除数据表之前应判断该表是否存在,避免报错。

DROP TABLE 可以同时删除多个表,只要将表名依次写在后面,相互之间用逗号隔开即可。

4.4 用户管理

MySQL 是一个多用户数据库,具有功能强大的访问控制系统,可以为不同用户指定允许的权限。MySQL 权限可以分为普通用户和 root 用户。root 用户是超级管理员,拥有所有权限,包括创建用户、删除用户和修改用户的密码等管理权限。普通用户只拥有被授予的各种权限,用户管理包括管理用户账户、权限等。

本节内容涉及到了数据库的安全,是数据库管理中非常重要的内容。学习本章可以学会有效保证 MySQL 数据库的安全。

4.4.1 权限表

MySQL 服务器通过权限表来控制用户对数据库的访问,权限表存放在 MySQL 数据库中,由 MySQL_install_db 进行脚本初始化。存储账户权限表主要有:user、db、tables_priv、columns_priv、procs_priv 等。

1. user 表

MySQL 在安装时会自动创建一个名为 mysql 数据库,mysql 数据库中存储的都是用户权限表。用户登录以后,MySQL 会根据这些权限表的内容为每个用户赋予相应的权限。

user 表是 MySQL 中最重要的一个权限表,用来记录允许连接到服务器的账号信息。需要注意的是,在 user 表里启用的所有权限都是全局级的,适用于所有数据库。

user 表中的字段大致可以分为 4 类,分别是用户列、权限列、安全列和资源控制列,下面主要介绍这些字段的含义。

(1)用户列

用户列存储了用户连接 MySQL 数据库时需要输入的信息。需要注意的是,从 MySQL 5.7 版本及之后不再使用 Password 来作为密码的字段,而改成了 authentication_string,见表 4-6 所列。

用户登录时,如果这 3 个字段同时匹配,MySQL 数据库系统才会允许其登录。创建新用户时,也是设置这 3 个字段的值。修改用户密码时,实际就是修改 user 表的 authentication_string 字段的值。因此,这 3 个字段决定了用户能否登录。

(2)权限列

权限列的字段决定了用户的权限,用来描述在全局范围内允许对数据和数据库进行的操作。权限大致分为两大类,分别是高级管理权限和普通权限:

表4-6 MySQL8.0版本的use表的用户列

字段名	字段类型	是否为NULL	默认值	说明
Host	char(255)	NO	无	主机名
User	char(32)	NO	无	用户名
authentication_string	text	YES	无	密码

①高级管理权限主要对数据库进行管理,例如关闭服务的权限、超级权限和加载用户等。

②普通权限主要操作数据库,例如查询权限、修改权限等。

user 表中对应的权限是针对所有用户数据库的,权限列包括 Select_priv、Insert_priv 等以 priv 结尾的字段。这些字段值的类型为 ENUM,可以取的值只能为 Y 和 N,Y 表示该用户有对应的权限,N 表示该用户没有对应的权限。查看 user 表的结构可以看到,这些字段的值默认都是 N。如果要修改权限,可以使用 GRANT 语句或 UPDATE 语句更改 user 表的这些字段来修改用户对应的权限。

如果要修改权限,可以使用 GRANT 语句为用户赋予一些权限,也可以通过 UPDATE 语句更新 user 表的方式来设置权限,后续权限管理会详细说明。

(3)安全列

安全列主要用来判断用户是否能够登录成功,MySQL8.0 版本安全列主要有 6 个字段,其中两个是 ssl 相关的,两个是 x509 相关的,另外两个是授权插件相关的。ssl 用于加密;x509 标准可用于标识用户;Plugin 字段标识是可以用于验证用户身份的插件,如果该字段为空,服务器会使用内建授权验证机制验证用户身份。可以通过 SHOW VARIABLES LIKE 'have_openssl'语句来查询服务器是否支持 ssl 功能。

(4)资源控制列

资源控制列的字段用来限制用户使用的资源,user 表中的资源控制列见表 4-7 所列。

表4-7 user 表的资源控制列

字段名	字段类型	是否为NULL	默认值	说明
max_questions	int unsigned	非空	0	规定每小时允许执行查询的操作次数
max_updates	int unsigned	非空	0	规定每小时允许执行更新的操作次数
max_connections	int unsigned	非空	0	规定每小时允许执行的连接操作次数
max_user_connections	int unsigned	非空	0	规定允许同时建立的连接次数

2. db 表

db 表是 MySQL 数据库中非常重要的权限表。db 表中存储了用户对某个数据库的操作权限,表中存储了用户对某个数据库的操作权限。表中的字段大致可以分为两类,分别是用户列和权限列。

(1)用户列

db 表用户列有 3 个字段,分别是 Host、User、Db,标识从某个主机连接某个用户对某个数据库的操作权限,这 3 个字段的组合构成了 db 表的主键。

(2)权限列

db 表中的权限列和 user 表中的权限列大致相同,只是 user 表中的权限是针对所有数据库的,而 db 表中的权限只针对指定的数据库。其中 db 表中 create_routine_priv 和 alter_routine_priv 这两个字段表明用户是否有创建和修改存储过程的权限。如果希望用户只对某个数据库有操作权限,可以先将 user 表中对应的权限设置为 N,然后在 db 表中设置对应数据库的操作权限。

3. tables_priv 表和 columns_priv 表

tables_priv 表用来对单个表进行权限设置,columns_priv 表用来对单个数据列进行权限设置。tables_priv 表中 Table_priv 对表的操作权限,包括 Select、Insert、Update、Delete Create、Drop、Grant、References、Index、Alter、Create View、Show view、Trigger 等。Column_priv 字段表示对表中的列的操作权限,包括 Select、Insert、Update 和 References。columns_priv 表中 Column_priv 用来指定对哪些数据列具有操作权限。

4. procs_priv 表

procs_priv 表可以对存储过程和存储函数进行权限设置,其中 Proc_priv 列表示拥有的权限,包括 Execute、Alter Routine、Grant 3 种。

4.4.2 账户管理

MySQL 提供了许多语句用来管理用户账号,这些语句可以用来管理包括登录和退出 MySQL 服务器、创建用户、删除用户、密码管理和权限管理等内容。MySQL 数据库的安全性,需要通过账户管理来保证。

1. MySQL 登录与退出

用户可以通过 mysql 命令来登录 MySQL 服务器,接下来将详细介绍 MySQL 中登录和退出服务器的方法。

启动 MySQL 服务后,可以使用以下命令来登录如下:

mysql -h hostname|hostIP -P port -u username -p DatabaseName -e "SQL 语句"

参数如下:

-h:指定连接 MySQL 服务器的地址。可以用两种方式表示,hostname 为主机名,hostIP 为主机 IP 地址。

-P：指定连接 MySQL 服务器的端口号，port 为连接的端口号。MySQL 的默认端口号是 3306，因此如果不指定该参数，默认使用 3306 连接 MySQL 服务器

-u：指定连接 MySQL 服务器的用户名，username 为用户名

-p：提示输入密码，即提示 Enter password

DatabaseName：指定连接到 MySQL 服务器后，登录到哪一个数据库中。如果没有指定，默认为 mysql 数据库

-e：指定需要执行的 SQL 语句，登录 MySQL 服务器后执行这个 SQL 语句，然后退出 MySQL 服务器

注意：大写的 P 表示端口号，小写的 p 表示密码，-p 和密码之间一定不能有空格，其他的像 -u、-h、-P 之类的，是可以有空格的，也可以没有空格。

在 Linux 服务器如果有多个 MySQL 数据库，可以通过嵌套字的方式登录，如 mysql> mysql -uroot -p -S /usr/local/mysqls.sock

退出 MySQL 服务器的方式很简单，只要在命令行输入 EXIT 或 QUIT 即可。"\q"是 QUIT 的缩写，也可以用来退出 MySQL 服务。

2. 新建用户

MySQL 在安装时，会默认创建一个名为 root 的用户，该用户拥有超级权限，可以控制整个 MySQL 服务器。在对 MySQL 的日常管理和操作中，为了避免有人恶意使用 root 用户控制数据库，我们通常创建一些具有适当权限的用户，尽可能地不用或少用 root 用户登录系统，以此来确保数据的安全访问。使用 CREATE USER 语句必须拥有 MySQL 数据库的 INSERT 权限或全局 CREATE USER 权限。

MySQL 提供了以下 3 种方法创建用户：

（1）使用 CREATE USER 语句，MySQL8 必须先创建用户再授权。

（2）直接操作 MySQL 授权表，既在 mysql.user 表中添加用户。

（3）最好的方法是使用 GRANT 语句，MySQL5.7 直接授权可创建用户。

案例 1：使用 CREATE USER 创建一个用户，用户名是 test，密码是 test，主机名是 %

mysql> create usertest@'%' identified with mysql_native_password by 'test'

mysql> flush privileges

案例 2：使用 INSERT 语句创建名为 test 的用户，主机名是 localhost，密码也是 test

INSERT INTO mysql.user(Host, User, authentication_string, ssl_cipher x509_issuer, x509_subject) VALUES ('localhost', 'test', 'test', '', '', '')

mysql> flush privileges

注意：MySQL 5.7 的 user 表中的密码字段从 Password 变成了 authentication_string，如果你使用的是 MySQL 5.7 之前的版本，将 authentication_string 字段替换成 Password 即可。

案例3：使用 GRANT 创建名为 test1 的用户，主机名为 %，密码为 test1，该用户对所有数据库的所有表都有 SELECT 权限

mysql> grant SELECT ON *.* to test1@'%' identified by 'test1'
mysql> flush privileges

注意：其中，"*.*"表示所有数据库下的所有表，此命令仅适合 5.7 及之前的版本。

3. 修改用户

语法格式如下：

RENAME USER ＜旧用户＞ TO ＜新用户＞

语法：

＜旧用户＞：系统中已经存在的 MySQL 用户账号。
＜新用户＞：新的 MySQL 用户账号

使用 RENAME USER 语句时应注意以下几点：
(1) RENAME USER 语句用于对原有的 MySQL 用户进行重命名
(2) 若系统统中旧账户不存在或者新账户已存在，该语句执行时会出现错误
(3) 使用 RENAME USER 语句，必须拥有 mysql 数据库的 UPDATE 权限或全局 CREATE USER 权限

案例1：使用 RENAME USER 语句将用户名 jeames 修改为 it，主机是 localhost

mysql> RENAME USER 'jeames'@'localhost' TO 'it'@'localhost'

4. 删除用户

在 MySQL 数据库中，可以使用 DROP USER 语句删除用户，也可以直接在 mysql.user 表中删除用户以及相关权限。

使用 DROP USER 语句删除用户的语法格式如下：
DROP USER ＜用户1＞ [，＜用户2＞]…

使用 DROP USER 语句应注意以下几点：
(1) 用户用来指定需要删除的用户账号。
(2) DROP USER 语句可用于删除一个或多个用户，并撤销其权限。
(3) 使用 DROP USER 语句必须拥有 MySQL 数据库的 DELETE 权限或全局 CREATE USER 权限。
(4) 在 DROP USER 语句的使用中，若没有明确地给出账户的主机名，则该主机名默认为"%"。
(5) 用户的删除不会影响它们之前所创建的表、索引或其他数据库对象，因为 MySQL 并不会记录是谁创建了这些对象。

案例1：用 DROP USER 语句删除用户'it'@'localhost'

mysql> DROP USER 'it'@'localhost'

案例2：使用 DELETE 语句删除普通用户'test2'@'localhost'

mysql> DELETE FROM mysql. user WHERE Host= 'ocalhost' AND User= 'test2'

5. 用户密码修改

在使用数据库时，我们也许会遇到 MySQL 需要修改密码的情况，比如密码太简单，或是为了安全考虑等。

（1）SET PASSWORD 命令

步骤1：输入命令 mysql －u root －p 指定 root 用户登录 MySQL

步骤2：set password for username@'localhost' = password('newpwd')

mysql> set password for root@'%'= password('root')

mysql> flush privileges;

其中，username 为要修改密码的用户名，newpwd 为要修改的新密码

步骤3：输入 quit；命令退出 MySQL 重新登录，输入新密码"root"登录就可以了

注意：此种方法适用于 MySQL5.7 以下版本。

（2）mysqladmin 命令

使用 mysqladmin 命令修改 MySQL 的 root 用户密码格式为：

mysqladmin －u 用户名 －p 旧密码 password 新密码

mysql> mysqladmin - uroot - proot password test

mysql> flush privileges;

注意：- uroot 和 - proot 是整体，如果区分端口请加 - P，此种方法适用于 MySQL5.7 以下版本。

（3）UPDATE 修改 user 表

步骤1：输入命令 mysql －u root －p 指定 root 用户登录 MySQL

#5.7 版本命令

mysql> update mysql. user set authentication_string = password('root') where user= 'root'

mysql> flush privileges

#8.0 版本命令

mysql> alter user root@'localhost' identified with mysql_native_password by 'root'

mysql> flush privileges

6. 忘记 root 密码重置

对于 root 用户丢失密码这种特殊情况，MySQL 实现了对应的处理机制。大家可以通过特殊方法登录到 MySQL 服务器，然后在 root 用户下重新设置密码。

方法 1：以下环境为 Linux 系统

（1）用命令编辑/etc/my.cnf 配置文件，即：vim /etc/my.cnf 或者 vi /etc/my.cnf

（2）在[mysqld]下添加 skip-grant-tables，然后保存并退出

（3）重启 mysql 服务：service mysqld restart

（4）更改 root 用户名

重启以后，执行 mysql 命令进入 mysql 命令行

（5）修改 root 用户密码，此处注意，有时候会报不允许修改，先 flush privileges 再执行即可

mysql> alter user root@'localhost' identified with mysql_native_password by 'root'

mysql> flush privileges

注意：在这里，还有一种办法，只能去修改 mysql 的 user 表，将加密字段 authentication_string 置空，但不能修改 authentication_string 为其他值，然后使用空密码登录。

（6）把/etc/my.cnf 中的 skip-grant-tables 注释掉，然后重启 mysql，即：service mysqld restart

好了，下面就可以用 root 新的密码登录了！

方法 2：以下环境为 windows 环境二进制版本

（1）先关闭 MySQL 服务，打开 cmd 窗口

（2）G:\mysql-8.0.23-winx64\bin\mysqld --datadir=G:\mysql-8.0.23-winx64\data80323308 --console --skip-grant-tables --shared-memory

shared-memory 表示以内存方式启动

（3）然后再开一个窗口，进入 bin 目录，登录 MySQL，空密码登录

cd G:\mysql-8.0.23-winx64\bin

G:\mysql-8.0.23-winx64\bin> mysql -uroot -p

（4）修改 root 用户，创建远程用户

mysql> flush privileges;

mysql> alter user root@'localhost' identified with mysql_native_password by 'root'

mysql> create user root@'localhost' identified with mysql_native_password by 'root'

mysql> grant all on *.* to root@'localhost' with grant option

mysql> flush privileges

7. 故障案例:MySQL8.0.26修改简易密码root报错处理

alter user root@'localhost' identified with mysql_native_password by 'root';
ERROR 1819 (HY000): Your password does not satisfy the current policy requirements

(1)查询密码策略,如图4-1所示。

mysql> SHOW VARIABLES LIKE 'validate_password%'

```
mysql> SHOW VARIABLES LIKE 'validate_password%';
+--------------------------------------+--------+
| Variable_name                        | Value  |
+--------------------------------------+--------+
| validate_password.check_user_name    | ON     |
| validate_password.dictionary_file    |        |
| validate_password.length             | 8      |
| validate_password.mixed_case_count   | 1      |
| validate_password.number_count       | 1      |
| validate_password.policy             | MEDIUM |
| validate_password.special_char_count | 1      |
+--------------------------------------+--------+
```

图4-1 MySQL8密码策略

(2)去除密码验证策略。

--设置为ON时可以将密码设置成当前用户名,默认关闭

mysql> set global validate_password.check_user_name= ON

--密码强度检查等级,0/LOW、1/MEDIUM、2/STRONG。默认是1,即MEDIUM

mysql> set global validate_password_policy= 0

mysql> set global validate_password.length= 4

--特殊字符

mysql> set global validate_password.mixed_case_count= 0

mysql> set global validate_password.number_count= 0

mysql> flush privileges

此时修改为密码root则OK,因为已经去除密码策略

密码验证策略

0/LOW:只检查长度

1/MEDIUM:检查长度、数字、大小写、特殊字符

2/STRONG:检查长度、数字、大小写、特殊字符字典文件

4.4.3 权限管理

权限管理主要是对登录到MySQL的用户进行权限验证。所有用户的权限都存储在MySQL的权限表中,不合理的权限规划会给MySQL服务器带来安全隐患。

数据库管理员要对所有用户权限进行合理规划管理。

1. 用户授权

权限信息被存储在 MySQL 数据库的 user、db、host、tables_priv、columns_priv 和 procs_priv 表中。在 MySQL 启动时,服务器会将这些数据库表中的权限信息的内容读入内存。

授权就是为某个用户授予权限,合理的授权可以保证数据库的安全。MySQL 中可以使用 GRANT 语句为用户授予权限。

在 MySQL 中,拥有 GRANT 权限的用户才可以执行 GRANT 语句,其语法格式如下:
```
GRANT priv_type [(column_list)] ON database.table
TO user [IDENTIFIED BY [PASSWORD] 'password']
[, user[IDENTIFIED BY [PASSWORD] 'password']]...
[WITH with_option [with_option]...]
```

语法:

(1) priv_type 参数表示权限类型;

(2) columns_list 参数表示权限作用于哪些列上,省略该参数时,表示作用于整个表。

(3) database.table 用于指定权限的级别。

(4) user 参数表示用户账户,由用户名和主机名构成,格式是"'username'@'hostname'"。

(5) IDENTIFIED BY 参数用来为用户设置密码。

(6) password 参数是用户的新密码。

WITH 关键字后面带有一个或多个 with_option 参数。这个参数有 5 个选项,详细介绍如下:

(7) GRANT OPTION:被授权的用户可以将这些权限赋予给别的用户。

(8) MAX_QUERIES_PER_HOUR count:设置每小时可以允许执行 count 次查询。

(9) MAX_UPDATES_PER_HOUR count:设置每小时可以允许执行 count 次更新。

(10) MAX_CONNECTIONS_PER_HOUR count:设置每时可以建立 count 个连接。

(11) MAX_USER_CONNECTIONS count:设置单个用户可以同时具有的 count 个连接。

语法:

MySQL 中可以授予的权限有如下几组:

列权限,和表中的具体列相关。

表权限,和具体表中的所有数据相关。

数据库权限,和具体的数据库中的所有表相关。

用户权限,和 MySQL 中所有的数据库相关。

对应地,在 GRANT 语句中可用于指定权限级别的值有以下几类格式
(1) * 表示当前数据库中的所有表。
(2) *.* 表示所有数据库中的所有表。
(3) db_name.* 表示某个数据库中的所有表,db_name 指定数据库名。
(4) db_name.tbl_name 表示某个数据库中的某个表或视图,db_name 指定数据库名,tbl_name 指定表名或视图名。
(5) db_name.routine_name 表示某个数据库中的某个存储过程或函数,routine_name 指定存储过程名或函数名。
(6) TO 子句表示如果权限被授予给一个不存在的用户,MySQL 会自动执行一条 CREATE USER 语句来创建这个用户,但同时必须为该用户设置密码。

数据库管理员给普通用户授权时一定要特别小心,如果授权不当,可能会给数据库带来致命的破坏。一旦发现给用户的权限太多,应该尽快使用 REVOKE 语句将权限收回。此处特别注意,最好不要授予普通用户 SUPER 权限,GRANT 权限。

案例1:root 用户给普通用户授予操作数据库表的权限

mysql> grant select,insert,update,delete on db1. student to jeames@'%'

案例2:创建、修改、删除数据库表结构的权限

mysql> grant create on db1.* to 'jeames'@'%'

mysql> grant alter on db1.* to 'jeames'@'%'

mysql> grant drop on db1.* to 'jeames'@'%'

案例3:操作数据库存储过程、函数的权限

mysql> grant create routine on db1.* to 'jeames'@'%'

mysql> grant alter routine on db1.* to 'jeames'@'%'

mysql> grant execute on db1.* to 'jeames'@'%'

2. 收回权限

收回权限就是取消已经赋予用户的某些权限。收回用户不必要的权限在一定程度上可保证系统的安全性。MySQL 中使用 REVOKE 语句取消用户的某些权限。使用 REVOKE 收回权限之后,用户账户的记录将从 db、tables_priv 和 column_priv 表中删除,但是用户账号记录仍然在 user 表中保存。要使用 REVOKE 语句,必须拥有 MySQL 数据库的全局 CREATE USER 权限或 UPDATE 权限。

第一种:删除用户某些特定的权限,语法格式如下:

REVOKE priv_type [(column_list)]...
ON database. table
FROM user [, user]...

REVOKE 语句中的参数与 GRANT 语句的参数意思相同。其中:

priv_type 参数表示权限的类型

column_list 参数表示权限作用于哪些列上,没有该参数时作用于整个表上
user 参数由用户名和主机名构成,格式为"username'@'hostname'"

案例:回收权限,谁授权谁回收

mysql> revoke select,insert,update,delete on db1.student from 'jeames'@'localhost'

第二种:收回特定用户的所有权限,语法格式如下:
REVOKE ALL PRIVILEGES, GRANT OPTION FROM user [, user]...

案例:使用 REVOKE 语句取消用户 jeames'的所有权限

mysql> REVOKE ALL PRIVILEGES, GRANT OPTION FROM 'jeames'@'localhost'

3. 查看权限

在 MySQL 中,可以通过查看 mysql.user 表中的数据记录来查看相应的用户权限,也可以使用 SHOW GRANTS 语句查询用户的权限。

第一种方法:查看 mysqluser 表

mysql> SELECT privileges_list FROM user WHERE user= 'username',host = 'hostname'

其中,privileges_list 为想要查看的权限字段,可以为 Select_priv、Insert_priv 等。

第二种方法:SHOW GRANTS 语句查询

GRANT 可以显示更加详细的权限信息,包括全局级的和非全局级的权限,如果表层级和列层级的权限被授予用户,它们也能在结果中显示出来。

mysql> SHOW GRANTS FOR 'username'@'hostname'
案例: mysql> SHOW GRANTS FOR 'user1'@'localhost'

```
+-------------------------------------+
| Grants for user1@localhost          |
+-------------------------------------+
| GRANT USAGE ON *.*  TO 'user1'@'localhost'|
+-------------------------------------+
1 row in set (0.00 sec)
```

4.5 SQL 实战

SQL 对于现在的互联网公司产研岗位、数据分析、数据库运维岗几乎是必备技能。如果你是数据分析师,需要熟练地把自己脑子里的数据和指标需求翻译成 SQL 逻辑去查询数据,进而完成自己的数据分析报告等,你的产出是分析报告,而

不是SQL代码。如果你是数仓工程师（偏应用层），需要根据业务逻辑去设计模型，编写调度任务去产出数据，以供业务人员使用，你的产出是数据模型和表。如果你是算法工程师，可能需要用SQL来实现用户标签、特征工程等工作，但是这些是为你的模型训练评估做基础准备工作，你的产出可以提升某些指标的算法模型。

所以，SQL每个人都要用，但是用来衡量产出的并不是SQL本身，你需要用这个工具，去创造其他的价值。

4.5.1 什么是SQL？

SQL是由IBM公司在1974～1979年之间根据E.J.Codd发表的关系数据库理论为基础开发的，其前身是"SEQUEL"，后更名为SQL。由于SQL语言集数据查询、数据操纵、数据定义和数据控制功能于一体，类似自然语言、简单易用以及非过程化等特点，得到了快速的发展，并于1986年10月，被美国国家标准协会（American National Standards Institute, ANSI）采用为关系数据库管理系统的标准语言，后为国际标准化组织（International Organization for Standardization, ISO）采纳为国际标准。

4.5.2 SQL分类

SQL(Structured Query Language)是结构化查询语言的简称，它是一种数据库查询和程序设计语言，同时也是目前使用最广泛的关系型数据库操作语言。在数据库管理系统中，使用SQL语言来实现数据的存取、查询、更新等功能。SQL是一种非过程化语言，只需要提出"做什么"，而不需要指明"怎么做"。

如图4-2所示，SQL语言分为五个部分，本节重点介绍DQL及DML语言。

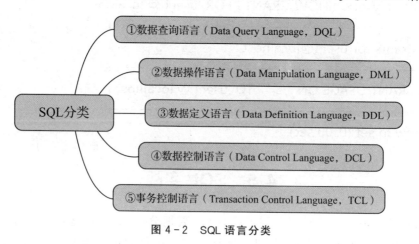

图4-2 SQL语言分类

(1) 数据查询语言(Data Query Language,DQL)

DQL 主要用于数据的查询,其基本结构是使用 SELECT 子句

FROM 子句和 WHERE 子句的组合来查询一条或多条数据

(2) 数据操作语言(Data Manipulation Language,DML)

DML 主要用于对数据库中的数据进行增加、修改和删除的操作,其主要包括:

INSERT:增加数据

UPDATE:修改数据

DELETE:删除数据

(3) 数据定义语言(Data Definition Language,DDL)

DDL 主要用针对是数据库对象(表、索引、视图、＞触发器、存储过程、函数、表空间等)进行创建、修改和删除操作。其主要包括:

CREATE:创建数据库对象

ALERT:修改数据库对象

DROP:删除数据库对象

(4) 数据控制语言(Data Control Language,DCL)

DCL 用来授予或回收访问数据库的权限,其主要包括:

GRANT:授予用户某种权限

REVOKE:回收授予的某种权限

事务控制语言(Transaction Control Language,TCL):

(5) TCL 用于数据库的事务管理

START TRANSACTION:开启事务

COMMIT:提交事务

ROLLBACK:回滚事务

SET TRANSACTION:设置事务的属性

4.5.3　DQL 数据查询

数据查询不应只是简单查询数据库中存储的数据,还应该根据需要对数据库进行筛选,同时进行一些逻辑运算,最终确定数据以什么样的格式显示。

MySQL 提供了功能强大、灵活的语句来实现这些操作。

MySQL 从数据表中查询数据的基本语句为 SELECT 语句,SELECT 语句的基本格式为:

SELECT

　　{*|<字段列表>}

　　[

　　　　FROM 表 1,表 2,...

```
    WHERE <表达式>
    [GROUP BY <group by definition>]
    [HAVING <expression> [<operator> <expression>]]
    [ORDER BY <order by definition>]
    [LIMIT [<offset>,] <row count>]
]
```

精简版格式：

SELECT * | 列名 FROM 表 WHERE 条件

(1){*|<字段列表>}包含星号通配符选择字段列表,至少包含一个字段名称,多个字段之间用逗号隔开。

(2)FROM<表1>,<表2>...,表1和表2表示查询数据的来源,也可以为视图。

(3)WHERE 子句是可选项,如果选择该项,将限定查询行必须满足的查询条件。

(4)GROUP BY<字段>,按照指定的字段分组,一般用于汇总分析。

(5)[ORDER BY<字段>],排序分升序(ASC)及降序(DESC)。

(6)[LIMIT[<offset>,]<row count>],显示查询出来的数据条数。

1. 简单查询

(1)检索数据

检索是指从一张表数据中查询所需的数据,主要查询方式有查询所有字段及查询指定字段,Customers 表结构及数据如图 4-3 所示。

```
mysql> desc Customers;
+-----------+--------------+------+-----+---------+-------+
| Field     | Type         | Null | Key | Default | Extra |
+-----------+--------------+------+-----+---------+-------+
| cust_id   | varchar(255) | NO   |     | NULL    |       |
| cust_name | varchar(255) | NO   |     | NULL    |       |
+-----------+--------------+------+-----+---------+-------+
2 rows in set (0.00 sec)

mysql> select * from Customers;
+---------+-----------+
| cust_id | cust_name |
+---------+-----------+
| a1      | andy      |
| a2      | ben       |
| a3      | tony      |
| a4      | tom       |
| a5      | an        |
| a6      | lee       |
| a7      | hex       |
+---------+-----------+
7 rows in set (0.00 sec)
```

图 4-3 Customers 表结构及数据

案例 1：现在有 Customers 表,返回所有列

mysql> select * from Customers

案例 2：现在有 Customers 表，只返回客户姓名（cust_name）列

mysql> select cust_name from Customers

（2）别名查询

执行查询时，为了方便操作或者需要多次使用相同的表时，可以为表指定别名，同时列的名称会很长或名称不够直观的时候，可以指定列别名，替换字段或表达式。Vendors 表结构及数据如图 4-4 所示。

```
mysql> desc Vendors;
+--------------+--------------+------+-----+---------+-------+
| Field        | Type         | Null | Key | Default | Extra |
+--------------+--------------+------+-----+---------+-------+
| vend_id      | varchar(255) | NO   |     | NULL    |       |
| vend_name    | varchar(255) | NO   |     | NULL    |       |
| vend_address | varchar(255) | NO   |     | NULL    |       |
| vend_city    | varchar(255) | NO   |     | NULL    |       |
+--------------+--------------+------+-----+---------+-------+
4 rows in set (0.01 sec)

mysql> select * from Vendors;
+---------+---------------+--------------+-----------+
| vend_id | vend_name     | vend_address | vend_city |
+---------+---------------+--------------+-----------+
| a001    | tencent cloud | address1     | shenzhen  |
| a002    | huawei cloud  | address2     | dongguan  |
| a003    | aliyun cloud  | address3     | alibaba   |
+---------+---------------+--------------+-----------+
3 rows in set (0.00 sec)
```

图 4-4　Vendors 表结构及数据

案例 1：编写 SQL 语句，从 Vendors 表中检索 vend_id、vend_name、vend_address 和 vend_city，将 vend_name 重命名为 vname，将 vend_city 重命名为 vcity，将 vend_address 重命名为 vaddress

mysql> select vend_id,vend_name as vname,vend_address as vaddress,vend_city as vcity from Vendors

别名的常见用法是在检索出的结果中重命名表的列字段。

表别名，一般用于多表查询中。

列使用别名后，如果做排序，一定要用别名。

可以用双引号或反引号将别名包围起来。

（3）去重查询

出于对数据分析的要求，需要消除重复的记录值，可以使用 DISTINCT 关键字指示 MySQL 消除重复的记录值。OrderItems 表结构及数据如图 4-5 所示。

案例 1：现在有 OrderItems 表，编写 SQL 语句，检索，并列出所有已订购商品（prod_id）的去重后的清单。

mysql> select distinct prod_id from OrderItems

```
mysql> desc OrderItems;
+---------+--------------+------+-----+---------+-------+
| Field   | Type         | Null | Key | Default | Extra |
+---------+--------------+------+-----+---------+-------+
| prod_id | varchar(255) | NO   |     | NULL    |       |
+---------+--------------+------+-----+---------+-------+
1 row in set (0.01 sec)

mysql> select * from OrderItems;
+---------+
| prod_id |
+---------+
| a1      |
| a2      |
| a3      |
| a4      |
| a5      |
| a6      |
| a6      |
+---------+
7 rows in set (0.00 sec)
```

图 4-5　OrderItems 表结构及数据

(4)运算查询

通过 MySQL 运算符进行运算,就可以获取表结构以外的另一种数据。常用的算数运算符包括加法、减法、乘法、除法及求余等。Products 表结构及数据如图 4-6 所示。

```
mysql> desc Products;
+------------+--------------+------+-----+---------+-------+
| Field      | Type         | Null | Key | Default | Extra |
+------------+--------------+------+-----+---------+-------+
| prod_id    | varchar(255) | NO   |     | NULL    |       |
| prod_price | double       | NO   |     | NULL    |       |
+------------+--------------+------+-----+---------+-------+
2 rows in set (0.00 sec)

mysql> select * from Products;
+---------+------------+
| prod_id | prod_price |
+---------+------------+
| a0011   |       9.49 |
| a0019   |        600 |
| b0019   |       1000 |
+---------+------------+
3 rows in set (0.00 sec)
```

图 4-6　Products 表结构及数据

案例 1:从 Products 表中返回的 prod_id、prod_price 和 sale_price。sale_price 是包含促销价格的计算字段

促销价格乘以 0.8,得到原价的 80%(即 20%的折扣)。

mysql> select prod_id,prod_price,prod_price* 0.8 sale_price from Products;

第 4 章　DBMS 基本管理

（5）数据过滤

根据特殊要求，只需要查询表中的指定数据，即对数据进行过滤。在 SELECT 语句中，通过 WHERE 子句可以对数据进行过滤，数据过滤一般会用到比较运算符，常用的比较和逻辑运算符见表 4-8 所列。

表 4-8　常用比较和逻辑运算符

比较运算符	说明
=	相等
<>,!=	不相等
<	小于
<=	小于或者等于
>	大于
>=	大于或者等于
BETWEEN	位于两值之间
IS NULL	判断一个值是否为 NULL
IS NOT NULL	判断一个值是否不为 NULL
IN	判断一个值是 IN 列表中的任意一个值
NOT IN	判断一个值不是 IN 列表中的任意一个值
LIKE	通配符匹配
REGEXP	正则表达式匹配

案例 1：编写 SQL 语句，返回 Products 表中所有价格在 3～6 美元之间的产品的名称（prod_name）和价格（prod_price），然后按价格对结果进行排序。

mysql> select prod_name,prod_price from Products where prod_price between 3 and 6 order by prod_price

案例 2：列举常用的数据过滤

①查询 score 表中成绩在 50～80 之间的所有行（区间查询和运算符查询）

mysql> SELECT * FROM score WHERE degree BETWEEN 50 AND 80

mysql> SELECT * FROM score WHERE degree >= 60 AND degree <= 80

②查询 student 表中'95033'班或性别为'女'的所有行

mysql> SELECT * FROM student WHERE class = '95033' or sex = '女'

③查询第二个字为'蔻'的所有商品

mysql> SELECT * FROM product WHERE pname like '_蔻%'

④查询 category_id 为 NULL 的商品

mysql> SELECT * FROM product WHERE category_id IS NULL

⑤查询价格不是 800 的所有商品

mysql> SELECT * FROM product WHERE NOT(price = 800)

案例 3：Jeames 想要找出 goods 表中所有名称包含牛奶的冰激凌，应该怎么写这个查询

mysql> select * from goods where name regexp '牛奶.*冰激凌'

说明：

MySQL 中匹配正则表达式需要使用关键字 REGEXP，在 REGEXP 关键字后面跟上正则表达式的规则即可。

BETWEEN AND 操作符前可以加关键字 NOT，表示指定范围之外的值。

可以和 LIKE 一起使用的通配符有'％'和'_'。

百分号通配符'％'，匹配任意长度的字符，甚至包括零字符。

下画线通配符'_'，一次只能匹配任意一个字符。

空值不同于 0，也不同于空字符串，空值一般表示数据未知。

（6）排序查询

MySQL 可以通过在 SELECT 语句中使用 ORDER BY 子句，对查询结果进行排序。student 表结构及数据如图 4-7 所示。

```
mysql> desc student;
+---------+-------------+------+-----+---------+-------+
| Field   | Type        | Null | Key | Default | Extra |
+---------+-------------+------+-----+---------+-------+
| no      | varchar(20) | NO   | PRI | NULL    |       |
| name    | varchar(20) | NO   |     | NULL    |       |
| sex     | varchar(10) | NO   |     | NULL    |       |
| birthday| date        | YES  |     | NULL    |       |
| class   | varchar(20) | YES  |     | NULL    |       |
+---------+-------------+------+-----+---------+-------+
5 rows in set (0.01 sec)

mysql> select * from student;
+-----+--------+-----+------------+-------+
| no  | name   | sex | birthday   | class |
+-----+--------+-----+------------+-------+
| 101 | 曾华   | 男  | 1977-09-01 | 95033 |
| 102 | 匡明   | 男  | 1975-10-02 | 95031 |
| 103 | 王丽   | 女  | 1976-01-23 | 95033 |
| 104 | 李军   | 男  | 1976-02-20 | 95033 |
| 105 | 王芳   | 女  | 1975-02-10 | 95031 |
| 106 | 陆军   | 男  | 1974-06-03 | 95031 |
| 107 | 王飘飘 | 男  | 1976-02-20 | 95033 |
| 108 | 张全蛋 | 男  | 1975-02-10 | 95031 |
| 109 | 赵铁柱 | 男  | 1974-06-03 | 95031 |
+-----+--------+-----+------------+-------+
9 rows in set (0.00 sec)
```

图 4-7　student 表结构及数据

案例1：以class降序的方式查询student表的所有行

mysql> SELECT * FROM student ORDER BY class DESC

案例2：以class降序、birthday升序查询student表的所有行

mysql> SELECT * FROM student ORDER BY birthday ASC, class DESC

①asc代表升序,desc代表降序,不声明默认为升序
②order by 用于子句可以支持单个字段,多个字段,表达式,函数,别名
③order by 子句放在查询语句最后面,LIMIT子句除外
④常用LIMIT使用如下：
LIMIT r：表示前r条数据

mysql> SELECT * FROM student order by no desc limit 5

LIMIT r, n：表示从第r行开始,查询n条数据

mysql> SELECT * FROM student LIMIT 0, 6

LIMIT n offset r：表示查询n条数据,从第r行开始

mysql> SELECT * FROM student LIMIT 3 offset 2

(7)分组查询

分组查询是对数据按照某个或多个字段进行分组,MySQL中使用GROUP BY关键字对数据进行分组,使用HAVING过滤分组。GROUP BY关键字通常和聚合函数一起使用,常用聚合函数见表4-9所列。

表4-9 常用聚合函数

聚合函数	作用
AVG()	返回某列的平均值
COUNT()	返回某列的行数
MAX()	返回某列的最大值
MIN()	返回某列的最小值
SUM()	返回某列的和

案例：员工信息如下：

```
create table employee
(
    id      serial primary key
    name    varchar(256)
    dept    varchar(256)
```

```
    salary decimal(12,4)
)
```

①每个部门工资最高的员工的全部信息

```
mysql> select id, dept, max(salary) as salary, name from employee group by dept
```

②每月工资开支超过十万的部门

```
mysql> select dept from employee group by dept having sum(salary) > 100000
```

①分组之后的条件筛选使用 having 实现：
SELECT 字段1，字段2... FROM 表名 GROUP BY 分组字段 HAVING 分组条件。

②HAVING 在数据分组之后进行过滤来选择分组，而 WHERE 在分组之前用来选择记录，另外 WHERE 排出的分组不包括在分组中。

③group by 子句用来分组 where 子句的输出。

④having 子句用来从分组的结果中筛选行。

⑤某些情况下需要对分组进行排序，ORDER BY 用来对查询的记录排序，如果和 GROUP BY 一起使用可以完成对分组的排序。

2. 子查询

子查询指一个查询语句嵌套在另一个查询语句内部的查询，在 SELECT 中，要先计算子查询，子查询结果作为外层另一个查询的过滤条件，查询可以基于一个表或者多个表。

案例1:Employee 表结构如下，编写一个 SQL 查询，获取，并返回 Employee 表中第二高的薪水。如果不存在第二高的薪水，查询应该返回 null。

```
+-------------+------+
| Column Name | Type |
+-------------+------+
| id          | int  |
| salary      | int  |
+-------------+------+
```

id 是这个表的主键。

表的每一行包含员工的工资信息。

```
mysql> SELECT MAX(Salary)  FROM Employee
Where Salary < (SELECT MAX(Salary) FROM Employee)
```

案例 2:jeames 想要从员工表

create table employee(
id int primary key auto_increment
name varchar(256)
dept varchar(64)
salary decimal(12, 4)
)

构造一个员工列表,排除每个部门最高工资的员工,这个查询怎样写?
mysql> select id, name, dept, salary
from employee as o
where o. salary < any(select salary from employee as i where i. dept = o. dept)

说明:
(1)子查询常用的操作符有 ANY、ALL、IN、EXISTS 等。
(2)EXISTS 关键字可以和条件表达式一起使用,EXISTS 与 NOT EXISTS 的结果只取决于是否会返回行,而不取决于这些行的内容。
(3)IN 关键字进行子查询时,内层查询语句仅返回一个数据列。
(4)ANY 表示至少一个,ALL 表示符合 SQL 语句中的所有条件。

3. 多表连接

连接是关系数据库模型的主要特点,通过连接运算符可以实现多个表查询,当两个或多个表中存在相同意义的字段时,便可以通过这些字段对不同的表进行连接查询。

MySQL 六种关联查询如下:

交叉连接(CROSS JOIN)
内连接(INNER JOIN)
外连接(LEFT JOIN/RIGHT JOIN)
联合查询(UNION 与 UNION ALL)
全连接(FULL JOIN)
自连接(Self JOIN)

案例中使用的 card 表及 person 表结构及数据如图 4-8 所示。

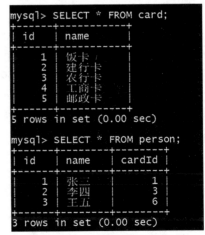

图 4-8　card 表及 person 表

内连接案例:

要查询这两张表中有关系的数据,可以使用 INNER JOIN (内连接)将它们连接在一起。

-- INNER JOIN:表示为内连接,将两张表拼接在一起

-- on：表示要执行某个条件
SELECT * FROM person INNER JOIN card on person.cardId = card.id
-- 将 INNER 关键字省略掉,结果也是一样的
mysql> SELECT * FROM person JOIN card on person.cardId = card.id

id	name	cardId	id	name
1	张三	1	1	饭卡
2	李四	3	3	农行卡

左外连接案例：

完整显示左边的表（person）,右边的表如果符合条件就显示,不符合则补 NULL。
LEFT JOIN 也叫作 LEFT OUTER JOIN,用这两种方式的查询结果是一样的。
mysql> SELECT * FROM person LEFT JOIN card on person.cardId = card.id

id	name	cardId	id	name
1	张三	1	1	饭卡
2	李四	3	3	农行卡
3	王五	6	NULL	NULL

右外连接案例：

完整显示右边的表（card）,左边的表如果符合条件就显示,不符合则补 NULL。
mysql> SELECT * FROM person RIGHT JOIN card on person.cardId = card.id

id	name	cardId	id	name
1	张三	1	1	饭卡
NULL	NULL	NULL	2	建行卡
2	李四	3	3	农行卡
NULL	NULL	NULL	4	工商卡
NULL	NULL	NULL	5	邮政卡

全外连接案例：

完整显示两张表的全部数据，MySQL 全连接语法，使用 UNION 将两张表合并在一起。

```
mysql> SELECT * FROM person LEFT JOIN card on person.cardId = card.id
UNION
SELECT * FROM person RIGHT JOIN card on person.cardId = card.id
+----+------+-------+------+--------+
| id | name | cardId| id   | name   |
+----+------+-------+------+--------+
|  1 | 张三 |     1 |    1 | 饭卡   |
|  2 | 李四 |     3 |    3 | 农行卡 |
|  3 | 王五 |     6 | NULL | NULL   |
|NULL| NULL | NULL  |    2 | 建行卡 |
|NULL| NULL | NULL  |    4 | 工商卡 |
|NULL| NULL | NULL  |    5 | 邮政卡 |
+----+------+-------+------+--------+
```

说明：

（1）UNION 和 UNION ALL 的区别：使用 UNION ALL 的功能是不删除重复行，而 UNION 去重。

（2）自连接是一种特殊的内连接，它是指相互连接的表在物理上为同一张表，但可以在逻辑上分为两张表。

（3）交叉连接查询其最终结果称为笛卡尔积，一般应尽量避免。

4.5.4 DML 数据操纵

MySQL 提供了功能丰富的数据库管理语句，包括有效地向数据库插入数据的 INSERT 语句，更新数据的 UPDATE 语句以及当数据不再使用时删除数据的 DELETE 语句。

1. 插入 INSERT

在 MySQL 中，可使用 INSERT 语句向数据库表中插入新的数据记录。可以插入的方式有插入完整的记录、插入记录的一部分、插入多条记录、插入另一个查询的结果。

tb_class 的表结构如下：

```
+-----------+----------------+------+-----+---------+-------+
| Field     | Type           | Null | Key | Default | Extra |
+-----------+----------------+------+-----+---------+-------+
| classNo   | char(6)        | NO   | PRI | NULL    |       |
| className | varchar(20)    | NO   | UNI | NULL    |       |
| department| varchar(20)    | YES  |     | NULL    |       |
| grade     | enum('1','2','3','4') | YES  |     | NULL    |       |
| classNum  | tinyint        | YES  |     | NULL    |       |
+-----------+----------------+------+-----+---------+-------+
```

案例1：向 tb_class 表中插入一条记录

mysql> INSERT INTO tb_class(classNo,department,className) VALUES('AC1301','会计学院','会计13-1班');

案例2：向 tb_class 表中插入多条记录

mysql> INSERT INTO tb_class(classNo,department,className) VALUES('AC1302','会计学院','会计13-2班'),('CS1401','计算机学院','计算机14-1班')

说明：

(1)要保证每个插入值的类型和对应列的数据类型匹配，如果类型不同，将无法插入。

(2)为表的指定字段插入数据，就是在 INSERT 语句中只向部分字段插入值，而其他字段的值为表定义时的默认值。

(3)一条 INSERT 语句和一条 SELECT 语句组成的组合语句，即可快速从一个表或多个表向一个表中插入多个行 INSERT INTO 表1(字段名列表)SELECT(字段名列表)FROM 表2 WHERE 查询条件。

2. 更新 UPDATE

MySQL 中使用 UPDATE 语句更新表中的记录，可以更新特定的行或者同时更新所有的行。

精简版格式：基本语法结构如下：

UPDATE 表名 SET 字段1=值1,字段2=值2,......,字段n=值n WHERE 条件

案例1：在 tb_class 表中，将 department 字段计算机学院替换为人工智能学院

mysql>update tb_class set department='人工智能学院' where department='计算机学院'

3. 删除 DELETE

从数据表中删除数据使用 DELETE 语句，DELETE 语句允许 WHERE 子句指定删除条件。

DELETE 语句基本语法格式如下：
DELETE FROM 表名［WHERE 条件］

注意：

从数据表中删除数据还可以使用 truncate，truncate table 命令，可快速删除数据表中的所有记录，truncate 操作会重置高水位线。delete 是可以带 WHERE 的，所以支持条件删除；而 truncate 只能删除整个表。

案例1：删除 tb_class 表中，className 为会计 13-1 班的信息
mysql> delete from tb_class where className='会计 13-1 班'

4.6 函数的用法

函数表示对输入参数值返回一个具有特定关系的值，MySQL 提供了大量丰富的内置函数，日常数据的查询和操作时会经常用到各种函数。MySQ 中的函数主要包括数值型函数、字符串函数、日期和时间函数、条件判断函数、系统信息函数和加密函数等。在实际环境中，这些函数可能嵌套使用，使用方法要复杂很多，希望大家用到的时候要多注意各个参数的作用。

4.6.1 数值型函数

数值型函数主要是对数值型数据进行处理，得到我们想要的结果，具体使用方法见表 4-10 所列。

表 4-10 数值型函数

函数	描述	实例	输出
ABS(x)	返回 x 的绝对值	SELECT ABS(-1)	1
CEIL(x)	返回大于或等于 x 的最小整数	SELECT CEIL(1.5)	2
FLOOR(x)	返回小于或等于 x 的最大整数	SELECT FLOOR(2.5);	2
GREATEST(expr1, expr2, expr3, ...)	返回列表中的最大值	SELECT GREATEST(3, 12, 34, 8, 25)	34
LEAST(expr1, expr2, expr3, ...)	返回列表中的最小值	SELECT LEAST(3, 12, 34, 8, 25)	3
MOD(x,y)	返回 x 除以 y 以后的余数	SELECT MOD(5,2)	1
POW(x,y)	返回 x 的 y 次方	SELECT POW(2,3)	8
SQRT(x)	返回 x 的平方根	SELECT SQRT(36)	6

续表 4-10

函数	描述	实例	输出
ROUND(x,y)	返回数值 x 保留到小数点后 y 位的值,四舍五入	SELECT ROUND(1.23456,4)	1.2346
TRUNCATE(x,y)	返回数值 x 保留到小数点后 y 位的值,截位,不会四舍五入	SELECT TRUNCATE(1.23456,3)	1.234

4.6.2 字符串函数

字符串函数主要用来处理数据库的字符串数据,具体使用方法见表 4-11 所列。

表 4-11 字符串函数

函数	描述	实例	输出
LENGTH(str)	字符串的字符长度	SELECT LENGTH('数据库');	9
CHAR_LENGTH(str)	字符串 str 所包含的字符个数	SELECT CHAR_LENGTH('数据库')	3
CONCAT(s1,s2...,sn)	连接参数产生的字符串	SELECT CONCAT('MySQL','8.0')	MySQL8.0
CONCAT_WS(x,s1,s2...sn)	合并多个字符串,并添加分隔符	SELECT CONCAT_WS("-","SQL","Tutorial","fun")	SQL-Tutorial-fun
LEAST(expr1,expr2,expr3,...)	返回列表中的最小值	SELECT LEAST(3, 12, 34, 8, 25)	3
INSERT(s1,x,len,s2)	字符串 s2 替换 s1 的 x 位置开始长度为 len 的字符串	SELECT INSERT('Football',2,4,'Play')	FPlayall
LOWER(str)	将字符串中的字母转换为小写	SELECT LOWER('Green')	green
UPPER(str)	将字符串中的字母转换为大写	SELECT UPPER('Green')	GREEN
LEFT(str,x)	返回字符串 str 中最左边 x 个字符	SELECT LEFT('MySQL',2)	My
RIGHT(str,x)	返回字符串 str 中最右边 x 个字符	SELECT RIGHT('MySQL',3)	SQL
TRIM(str)	删除字符串左右两侧的空格	SELECT TRIM(' MySQL ')	MySQL

续表 4-11

函数	描述	实例	输出
REPLACE(s, s1, s2)	字符串 s 中的字符 s1 替换为字符 s2	SELECT REPLACE('abc','a','x')	xbc
SUBSTR(s, start, length)	返回从指定位置开始的指定长度的字符串	select SUBSTRING('computer', 3,4)	mput
REVERSE(str)	返回颠倒字符串 str 的结果	SELECT REVERSE('abc')	cba

LENGTH(str)返回值为字符串的字节长度,使用 utf8(UNICODE 的一种变长字符编码,又称万国码)编码字符集时,一个汉字是 3 个字节,一个数字或字母算一个字节。

4.6.3 日期函数

日期和时间函数主要用来处理日期和时间值,许多日期函数可以同时接受数和字符串类型的两种参数,具体使用方法见表 4-12 所列。

表 4-12 日期函数

函数	描述	实例	输出
CURDATE/CURRENT_DATE	返回当前系统的日期值	select CURDATE()	2022-08-28
CURTIME/CURRENT_TIME	返回当前系统的时间值	select CURTIME()	09:04:01
NOW/SYSDATE	连接参数产生的字符串	select now()	2022-08-28 09:14:38
YEAR	获取指定日期中的年份	select YEAR('2022-8-28')	2022
MONTH	获取指定日期中的月份	select MONTH('2022-8-28')	8
WEEK	获取指定日期是一年中的第几周	select WEEK('2022-8-28')	35
DAYOFYEAR	指定日期是一年中的第几天	select DAYOFYEAR('2022-8-28')	240
DAYOFMONTH	指定日期是一个月中是第几天	select DAYOFMONTH('2022-8-28')	28
DATE	提取日期值	SELECT DATE('20170615')	2017-06-15

续表 4-12

函数	描述	实例	输出
ADDDATE(d,n)	起始日期 d 加上 n 天的日期	SELECT ADDDATE("2022-06-15", INTERVAL 10 DAY);	2022-06-25
DATEDIFF(d1,d2)	日期 d1->d2 之间相隔的天数	SELECT DATEDIFF('2022-08-01','2022-02-02')	180
DATE_FORMAT(d,f)	按表达式 f 的要求显示日期 d	SELECT DATE_FORMAT ('2022-08-11 11:11:11', '%Y-%m-%d %r')	2022-08-11 11:11:11 AM

DATE_ADD(d,INTERVAL expr type)表示计算起始日期 d 加上一个时间段后的日期,type 值可以是 MICROSECOND、SECOND、MINUTE、HOUR、DAY、WEEK、MONTH 等。

DATE_FORMAT(date,format) 函数是根据 format 指定的格式显示 date 值。

```
mysql> SELECT DATE_FORMAT('2022-08-08 21:45:00','%W %M %D %Y') AS col1
    DATE_FORMAT('2022-08-15 21:45:00','%h:i% %p %M %D %Y') AS col2
+----------------------+----------------------+
| col1                 | col2                 |
+----------------------+----------------------+
| Monday August 8th 2022 | 09:i PM August 15th 2022 |
+----------------------+----------------------+
```

4.6.4 条件判断函数

条件判断函数也称控制流程函数,要根据满足的条件不同,执行相应的流程,用来实现 SQL 的条件逻辑,允许开发者将一些应用程序业务逻辑转换到数据库后台。MySQL 中进行条件判断的函数有 IF、IFNULL、NULLIF 和 CASE。

IF(expr,v1,v2): 如果表达式 expr 成立, 返回结果 v1; 否则, 返回结果 v2

IFNULL(v1,v2): 如果 v1 的值不为 NULL, 则返回 v1, 否则返回 v2

NULLIF(expr1, expr2): 比较两个字符串,如果字符串 expr1 与 expr2 相等 返回 NULL, 否则返回 expr1

CASE

CASE WHEN[test1] THEN [result1]…ELSE [default] END 如果 testN 是真,则返回 resultN, 否则返回 default

CASE [test] WHEN[val1] THEN [result]…ELSE [default]END 如果 test 和 valN 相等, 则返回 resultN, 否则返回 default

案例1: mysql> SELECT IF(1 > 0,'正确','错误')
-> 正确
案例2: mysql> SELECT IFNULL(null,'Hello Word')
-> Hello Word
案例3: mysql> SELECT NULLIF(66, 25)
-> 66
案例4: mysql> SELECT CASE WHEN 1> 0 THEN 'true' ELSE 'false' END;
-> true
案例5: SELECT CASE test WHEN 1 THEN 'one' WHEN 2 THEN 'two'ELSE 'more END;-> more

4.6.5 系统信息函数

MySQL中的系统信息有数据库的版本号、当前用户名和连接数、系统字符集、最后一个自动生成的ID值等,具体使用方法见表4-13所列。

表4-13 MySQL系统信息函数

函数	描述	实例	输出
VERSION()	返回数据库的版本号	SELECT VERSION()	8.0.27
CONNECTION_ID()	返回当前连接的次数	select CONNECTION_ID()	9
DATABASE()	返回当前数据库名	SELECT DATABASE()	test
CURRENT_USER()/USER()	返回当前用户	SELECT CURRENT_USER()	root@localhost
CONNECTION_ID()	返回唯一的连接ID	SELECT CONNECTION_ID()	9
LAST_INSERT_ID()	最后一个自动生成的ID值	select LAST_INSERT_ID();	3
COLLATION(str)	返回字符串str的字符排序方式	select COLLATION('tre');	utf8mb4_0900_ai_ci
CHARSET(str)	返回字符串str自变量的字符集	select CHARSET('str')	utf8mb4

4.6.6 加密函数

加密函数主要用于对字符串进行加密,加密与解密函数主要用于对数据库中的数据进行加密和解密处理,以防止数据被他人窃取。这些函数在保证数据库安全时

非常有用,常用的几个列举如下:

(1)MD5(str)为字符串算出一个 MD5 128 比特校验和。

说明:该值以 32 位十六进制数字的二进制字符串形式返回,若参数为 NULL,则会返回 NULL。

(2)SHA(str):加密算法比 MD5 更加安全。

(3)ENCODE(str,pswd_str):返回使用 pswd_str 作为加密密码加密 str。

4.7 约 束

约束是一种限制,它通过限制表中的数据,来确保数据的完整性和唯一性。比如有的数据是必填项,就像个人信息认证的时候,或者填注册信息的时候,手机号这种就不能空着,所以就有了非空约束;又有的数据比如用户的身份证号,不能跟其他人的一样,所以就需要使用唯一约束等。约束能够帮助数据库管理员更好地管理数据库,并且能够确保数据库中数据的正确性和有效性。可以在创建表时规定约束(通过 CREATE TABLE 语句),或者在表创建之后通过 ALTER TABLE 语句规定约束。

MySQL 中主要有 6 种约束:主键约束、外键约束、唯一约束、检查约束、非空约束和默认值约束。根据约束数据列的限制,约束可分为:单列约束(只约束一列)、多列约束(可约束多列),根据约束的作用范围,约束可分为:列级约束(只能作用在一个列上,跟在列的定义后面)、表级约束(可以作用在多个列上,不与列一起,而是单独定义)。

4.7.1 主键约束

在建立数据表的时候,一般情况下,为了方便更快地查找表中的记录,都会要求在表中设置一个"主键"。"主键"是表里面的一个特殊字段,其关键字是 primary key,这个字段能够唯一标识该表中的每条信息。

使用主键的时候需要注意以下几个点:

(1)主键约束相当于唯一约束+非空约束的组合,主键约束列不允许重复,也不允许出现空值。

(2)每个表最多只允许一个主键。

(3)当创建主键的约束时,系统默认会在所在的列和列组合上建立对应的唯一索引。

1. 建表时设置主键约束

(1)在定义字段的时候设置主键约束(列级约束)

语法格式:

<字段名> <数据类型> PRIMARY KEY

例:在数据库中创建学生信息数据表 stinfo,主键为 stid,SQL 语句以及运行结果如下:

```
create table stinfo(
    stid int(10) primary key
    name varchar(20)
    class varchar(10)
    age int(2)
)
mysql> desc stinfo
```

Field	Type	Null	Key	Default	Extra
stid	int	NO	PRI	NULL	
name	varchar(20)	YES		NULL	
class	varchar(10)	YES		NULL	
age	int	YES		NULL	

(2)在定义完所有字段之后创建(表级约束)

语法格式:

[CONSTRAINT <约束名>] PRIMARY KEY [字段名]

案例:在数据库中创建学生信息数据表 stinfo_bk,主键为 stid,SQL 语句如下:

```
create table stinfo_bk(
    stid int(10),
    name varchar(20),
    class varchar(10),
    age int(2),
    CONSTRAINT stid_pk PRIMARY KEY(stid)
);
```

(3)设置联合主键

所谓的联合主键,就是这个主键是由一张表中多个字段组成的,语法格式:

PRIMARY KEY [字段1,字段2,…,字段n]

案例:在数据库中创建雇员表 emp,联合主键为 name,deptId,SQL 语句以及运行结果如下:

```
create table emp(
    name varchar(20)
    deptId int
    salary double
    primary key(name,deptId)
)
```

```
mysql> desc emp
+--------+-------------+------+-----+---------+-------+
| Field  | Type        | Null | Key | Default | Extra |
+--------+-------------+------+-----+---------+-------+
| name   | varchar(20) | NO   | PRI | NULL    |       |
| deptId | int         | NO   | PRI | NULL    |       |
| salary | double      | YES  |     | NULL    |       |
+--------+-------------+------+-----+---------+-------+
```

注意：在设置联合主键的时候，不能在每个字段名后面直接声明主键约束

2. 在修改表的时候添加主键约束

如果在创建表的时候没有设置主键约束，还可以在修改表时进行添加，语法格式如下：

ALTER TABLE ＜数据表名＞ ADD PRIMARY KEY(＜字段名＞);

案例：雇员表 emp_bk 已创建，添加联合主键 name,deptId，SQL 语句以及运行结果如下：

--未创建主键前
```
mysql> desc emp_bk
+--------+-------------+------+-----+---------+-------+
| Field  | Type        | Null | Key | Default | Extra |
+--------+-------------+------+-----+---------+-------+
| name   | varchar(20) | YES  |     | NULL    |       |
| deptId | int         | YES  |     | NULL    |       |
| salary | double      | YES  |     | NULL    |       |
+--------+-------------+------+-----+---------+-------+
```

--创建主键后
```
mysql> alter table emp_bk add primary key(name,deptId)
mysql> desc emp_bk
```

```
+------+------------+------+-----+---------+-------+
| Field| Type       | Null | Key | Default | Extra |
+------+------------+------+-----+---------+-------+
| name | varchar(20)| NO   | PRI | NULL    |       |
|deptId| int        | NO   | PRI | NULL    |       |
|salary| double     | YES  |     | NULL    |       |
+------+------------+------+-----+---------+-------+
```

注意：设置成主键约束的字段不允许有空值。

3. 删除主键约束

一个表中不需要主键约束时，就需要从表中将其删除，语法格式如下：

mysql> alter table <数据表名> drop primary key;

4. 设置主键自增长

在 MySQL 里，当主键定义为自增长后，主键的值就不需要自己再输入数据了，而是由数据库系统根据定义自动赋值，每增加一条记录，主键就会自动根据设置的步长进行增长。

通过给字段添加 auto_increment 属性来实现主键自增长，语法格式：

字段名 数据类型 AUTO_INCREMENT

例：创建学生信息表 st_info，指定 st_id 字段自增

```
create table st_info(
        st_id int(10) primary key auto_increment
        name varchar(20) not null
        class varchar(10)
        gender varchar(4)
        age int(2)
)
mysql> desc st_info
+--------+------------+------+-----+---------+----------------+
| Field  | Type       | Null | Key | Default | Extra          |
+--------+------------+------+-----+---------+----------------+
| st_id  | int        | NO   | PRI | NULL    | auto_increment |
| name   | varchar(20)| NO   |     | NULL    |                |
| class  | varchar(10)| YES  |     | NULL    |                |
| gender | varchar(4) | YES  |     | NULL    |                |
| age    | int        | YES  |     | NULL    |                |
+--------+------------+------+-----+---------+----------------+
```

此时可以不用再手动插入 st_id 的数据。
```
insert into st_info(name,class,gender,age)
values('张三','计算机1班','女',18)
       ('四','计算机2班','女',18)
       ('王五','管理学1班','男',18)
```

```
mysql> select * from st_info
+------+------+-------------+--------+-----+
| st_id | name | class       | gender | age |
+------+------+-------------+--------+-----+
|    1 | 张三 | 计算机1班   | 女     |  18 |
|    2 | 李四 | 计算机2班   | 女     |  18 |
|    3 | 王五 | 管理学1班   | 男     |  18 |
+------+------+-------------+--------+-----+
```

(1) 如果在插入某一行数据产生了报错,则最终自增字段可能会出现不连续的情况。

(2) 一个表中只能有一个字段使用 auto_increment 约束,必须为主键。

(3) auto_increment 约束的字段只能是整数类型(TINYINT、SMALLINT、INT、BIGINT 等)。

(4) auto_increment 约束字段的最大值受该字段的数据类型约束,如果达到上限,会失效。

(5) 指定自增字段初始值可以在建表时指定,也可以在建表后修改。
create table 表名 (id int primary key auto_increment, name varchar(20))。
alter table 表名 auto_increment = 100。

(6) delete 和 truncate 在删除后自增列的变化。
delete 数据之后自动增长从断点开始。
truncate 数据之后自动增长从默认起始值开始。

(7) 如果有自增,先新增主键,再新增自增,此时要删除主键,必须先删除主键。

4.7.2 唯一约束

唯一约束(Unique)是指所有记录中字段的值不能重复出现。例如为 id 字段加上唯一性约束后,每条记录的 id 值都是唯一的,不能出现重复的情况,但可以为空。

通过给字段添加 Unique 属性来实现唯一约束,语法格式:
方式1:<字段名> <数据类型> unique
方式2:alter table 表名 add constraint 约束名 unique(列)

案例 1：创建表时指定
```
create table user_info (
 id int
 name varchar(20)
 telno varchar(20) unique -- 指定唯一约束
)
```

案例 2：创建表之后创建
```
create table user_bk (
  id int
  name varchar(20)
  telno varchar(20)
)
```
mysql> alter table user_bk add constraint tel_uk unique(telno);
mysql> desc user_bk

Field	Type	Null	Key	Default	Extra
id	int	YES		NULL	
name	varchar(20)	YES		NULL	
telno	varchar(20)	YES	UNI	NULL	

案例 3：删除 user_bk 表唯一约束 tel_uk
mysql> alter table user_bk drop constraint tel_uk

4.7.3 非空约束

非空约束用来约束表中的字段不能为空，其特点如下：

(1) 默认，所有的类型的值都可以是 NULL，包括 INT、FLOAT 等数据类型。

(2) 非空约束只能出现在表对象的列上，只能某个列单独限定非空，不能组合非空。

(3) 一个表可以有很多列都分别限定了非空，空字符串不等于 NULL。

案例 1：创建表时创建，book_info 表 name 字段添加非空约束
mysql> CREATE TABLE book_info(bookid int,name varchar(20) not null, author char(18))

案例 2：建表后,给 book_info 表 author 字段新增非空约束
```
mysql> alter table book_info modify author char(18) not null
mysql> desc book_info
+-------+-------------+------+-----+---------+-------+
| Field | Type        | Null | Key | Default | Extra |
+-------+-------------+------+-----+---------+-------+
| bookid| int         | YES  |     | NULL    |       |
| name  | varchar(20) | NO   |     | NULL    |       |
| author| char(18)    | NO   |     | NULL    |       |
+-------+-------------+------+-----+---------+-------+
```

案例 3：删除 book_info 表 name、author 字段的非空约束
#相当于修改某个非注解字段,该字段允许为空
```
mysql> alter table book_info modify name varchar(20) NULL
mysql> alter table book_info modify author char(18)
mysql> desc book_info
+-------+-------------+------+-----+---------+-------+
| Field | Type        | Null | Key | Default | Extra |
+-------+-------------+------+-----+---------+-------+
| bookid| int         | YES  |     | NULL    |       |
| name  | varchar(20) | YES  |     | NULL    |       |
| author| char(18)    | YES  |     | NULL    |       |
+-------+-------------+------+-----+---------+-------+
```

4.7.4　默认值约束

给某个字段/某列指定默认值,一旦设置默认值,在插入数据时,如果此字段没有显式赋值,则赋值为默认值,关键字为 DEFAULT。

MySQL 默认值约束用来指定某列的默认值,语法如下：
方式1：＜字段名＞＜数据类型＞ default ＜默认值＞
方式2：alter table 表名 modify 列名 类型 default 默认值

案例 1：创建表 stu_info,并给 department 字段添加默认值——城市学院
```
create table if not exists stu_info (
stuNo CHAR(10) not NULL primary key comment '学号',
stuName VARCHAR(10) NOT null comment '姓名',
nation VARCHAR(10) comment '民族',
```

```
department varchar(16) comment'院系' DEFAULT '城市学院'
)
```

案例 2:建 stu_info 表之后,并给 nation 字段添加默认值汉

```
mysql> alter table stu_info modify nation varchar(10) default '汉'
mysql> desc stu_info
```

Field	Type	Null	Key	Default	Extra
stuNo	char(10)	NO	PRI	NULL	
stuName	varchar(10)	NO		NULL	
nation	varchar(10)	YES		汉	
department	varchar(16)	YES		城市学院	

案例 3:删除 stu_info 表默认约束

```
mysql> ALTER TABLE stu_info ALTER COLUMN nation DROP DEFAULT;
mysql> alter table stu_info modify column department varchar(16) default null
mysql> desc stu_info
```

Field	Type	Null	Key	Default	Extra
stuNo	char(10)	NO	PRI	NULL	
stuName	varchar(10)	NO		NULL	
nation	varchar(10)	YES		NULL	
department	varchar(16)	YES		NULL	

4.7.5 外键约束

外键约束是表的一个特殊字段,一般会和主键约束一起使用,用来确保数据的一致性。对于两个具有关联关系的表来说,相关联字段中主键所在的表就是主表(父表),外键所在的表就是从表(子表)。

所以外键就是用来建立主表与从表的关联关系,为两个表的数据建立连接,约束两个表中数据的一致性和完整性。

在定义外键时需要遵守以下规则:

(1) 主表必须已经存在于数据库中,或者是当前正在创建的表,如果是后一种情况,则主表与从表是同一个表,这样的表称作自参照表,这种结构称作自参照完整性。

(2) 必须为主表定义主键,允许在外键中出现空值。

(3) 在主表的表名后面指定列名或列名的组合,这个列或列的组合必须是主表的主键。

(4) 外键中列的数目及类型必须和主表的主键中列的数目及类型相同。

1) 可以加入关键字 FOREIGN KEY 来指定外键,用 REFERENCES 来连接与主表的关系,语法如下:

方式1:新建表时添加外键

［CONSTRAINT］［外键约束名称］FOREIGN KEY（外键字段名）REFERENCES 主表名（主键字段名）

方式2:已有表添加外键

ALTER TABLE 从表 ADD［CONSTRAINT］［外键约束名称］FOREIGN KEY（外键字段名）

2) 外键一旦删除,就会解除主表和从表间的关联关系,语法如下:

ALTER TABLE ＜表名＞ DROP FOREIGN KEY ＜外键约束名＞

案例1:tb_class 表的创建语句及结构如下,按照要求建表 tb_class

```
CREATE TABLE tb_class (
  classNo CHAR(6) PRIMARY KEY NOT NULL
  className VARCHAR(20) NOT NULL,
  department VARCHAR(20)
  grade ENUM('1','2','3','4')
  classNum TINYINT
  constraint uq_class unique(className)
) engine= InnoDB
```

```
mysql>  desc tb_class
+-----------+--------------------+------+-----+---------+-------+
| Field     | Type               | Null | Key | Default | Extra |
+-----------+--------------------+------+-----+---------+-------+
| classNo   | char(6)            | NO   | PRI | NULL    |       |
| className | varchar(20)        | NO   | UNI | NULL    |       |
| department| varchar(20)        | YES  |     | NULL    |       |
| grade     | enum('1','2','3','4')| YES |     | NULL    |       |
| classNum  | tinyint            | YES  |     | NULL    |       |
+-----------+--------------------+------+-----+---------+-------+
```

建立 tb_student 到 tb_class 的外键约束(两个表相同含义的属性是 classNo,因此 classNo 是 tb_student 的外键),约束名为 fk_student,并定义相应的参照动作,更新操作为级联(cascade),删除操作为限制(restrict)。

```sql
CREATE TABLE tb_student (
  studentNo CHAR(10) NOT NULL
  studentName VARCHAR(10) NOT NULL
  sex CHAR(2)
  birthday DATE
  native VARCHAR(20)
  nation VARCHAR(20) default '汉'
  classNo CHAR(6)
  constraint fk_student FOREIGN KEY (classNo)
  references tb_class(classNo) on delete restrict on update cascade
) engine= InnoDB
```

mysql> desc tb_student

Field	Type	Null	Key	Default	Extra
studentNo	char(10)	NO		NULL	
studentName	varchar(10)	NO		NULL	
sex	char(2)	YES		NULL	
birthday	date	YES		NULL	
native	varchar(20)	YES		NULL	
nation	varchar(20)	YES		汉	
classNo	char(6)	YES	MUL	NULL	

Mysql 外键设置中的 CASCADE、NO ACTION、RESTRICT、SET NULL。

CASCADE:父表 delete、update 的时候,子表会 delete、update 掉关联记录。

SET NULL:父表 delete、update 的时候,子表会将关联记录的外键字段所在列设为 null,所以注意在设计子表时外键不能设为 not null。

RESTRICT:如果想要删除父表的记录,而在子表中有关联该父表的记录,则不允许删除父表中的记录。

NO ACTION:同 RESTRICT,也是首先先检查外键。

案例 2:创建 tb_course 表,建表后添加外键,priorCourse,外键名字为 fk_course,指向 courseNo

```
+------------+-------------+------+-----+---------+-------+
| Field      | Type        | Null | Key | Default | Extra |
+------------+-------------+------+-----+---------+-------+
| courseNo   | char(6)     | NO   | PRI | NULL    |       |
| courseName | varchar(20) | NO   | UNI | NULL    |       |
| credit     | decimal(3,1)| NO   |     | NULL    |       |
| courseHour | tinyint     | NO   |     | NULL    |       |
| term       | tinyint     | YES  |     | NULL    |       |
| priorCourse| char(6)     | YES  |     | NULL    |       |
+------------+-------------+------+-----+---------+-------+
```

CREATE TABLE tb_course (
  courseNo CHAR(6) NOT NULL primary key comment '课程号'
  courseName VARCHAR(20) unique not NULL comment '课程名'
  credit DECIMAL(3,1) not NULL comment '学分'
  courseHour TINYINT(2) not NULL comment '课时数'
  term TINYINT(2) comment '开课学期',
  priorCourse CHAR(6) comment '先修课程'
) engine= InnoDB

#主表与从表是同一个表，这样的表称作自参照表，这种结构称作自参照完整性
mysql> ALTER TABLE tb_course ADD CONSTRAINT
fk_score FOREIGN KEY(priorCourse) REFERENCES tb_course(courseNo)

案例3：删除表 tb_course 外键约束 fk_score
mysql> ALTER TABLE tb_course DROP FOREIGN KEY fk_score

4.7.6 CHECK 约束

MySQL 8.0 中可以使用 check 约束，8.0 以下版本不可用。MySQL 可以使用简单的表达式来实现 CHECK 约束，也允许使用复杂的表达式作为限定条件，例如在限定条件中加入子查询。

(1)创建表时设置检查约束的语法格式如下：
CHECK ＜表达式＞
其中，"表达式"指的就是 SQL 表达式，用于指定需要检查的限定条件
(2)在修改表时添加检查约束，语法格式如下：
ALTER TABLE＜数据表名＞ ADD CONSTRAINT ＜检查约束名＞ CHECK(＜检查约束＞)
(3)删除检查约束，语法格式如下：
ALTER TABLE ＜数据表名＞ DROP CONSTRAINT ＜检查约束名＞

案例1：创建 tb_emp 表，要求 salary 字段值大于 1000 且小于 10000，SQL 语句如下所示：

```
CREATE TABLE tb_emp
(id INT(11) PRIMARY KEY
name VARCHAR(25)
salary FLOAT
CHECK(salary> 1000 AND salary< 10000)
)
```

案例2：给 tb_emp 数据表，添加 CHECK 约束，要求 id 字段值大于 50，SQL 语句如下所示：

```
mysql>  ALTER TABLE tb_emp ADD CONSTRAINT check_id CHECK(id> 50)
```

案例3：删除 tb_emp 数据表 CHECK 约束 check_id，SQL 语句如下所示：

```
mysql>  ALTER TABLE tb_emp DROP CONSTRAINT check_id
```

4.8 存储过程

存储过程（Stored Procedure）是一组为了完成特定功能的 SQL 语句集，经编译后存储在数据库中，用户通过指定存储过程的名字，并给定参数（如果该存储过程带有参数）来调用执行它。存储过程是数据库中的一个重要功能，存储过程可以用来转换数据、数据迁移、制作报表，它类似于编程语言，一次执行成功，就可以随时被调用，避免了开发人员重复编写相同 SQL 语句的问题，完成指定的功能操作。

存储过程的优点如下：
(1)增强了 SQL 语句的灵活性。
(2)固定的业务模块化封装，较少开发重复性。
(3)执行速度很快。
(4)存储过程被作为一种安全机制，充分得到了利用。
存储过程大大提高数据库的处理速度，同时也可以提高数据库编程的灵活性。

4.8.1 存储过程的创建

编写存储过程并不是件简单的事情，不仅需要复杂的 SQL 语句，还要有创建存储过程的权限。但使用存储过程会简化操作，减少冗余的操作步骤。同时，还可以减少操作过程中的失误，提高效率，因此存储过程是非常有用的。

可以使用 CREATE PROCEDURE 语句创建存储过程，语法格式如下：
CREATE PROCEDURE <过程名>（[过程参数[,…]]）<过程体>
[过程参数[,…]] 格式

[IN | OUT | INOUT] <参数名> <类型>

例：输出当前数据库的用户个数

```
DELIMITER //
  CREATE PROCEDURE myproc(OUT s int)
    BEGIN
      SELECT COUNT(*) INTO s FROM mysql.user;
    END
    //
DELIMITER ;
```

注意：MySQL 服务器在处理时会以遇到的第一条 SQL 语句结尾处的分号作为整个程序的结束符，而不再去处理存储过程体中后面的 SQL 语句，为了解决这个问题，通常使用 DELIMITER 命令将结束命令修改为其他字符，语法格式如下：

```
#成功执行这条 SQL 语句后,命令、语句或程序的结束标志就换为:"//"
mysql> DELIMITER //
```

若希望换回默认的分号";"输入下列语句即可，DELIMITER 和分号";"之间一定要有一个空格

```
mysql> DELIMITER ;

#调用存储过程
mysql> call myproc(@s)
mysql> select @s
+-----+
| @s  |
+-----+
|  6  |
+-----+
```

4.8.2 存储过程类型

按照传参不同，可以将存储过程分为如下几类：IN 表示输入参数，OUT 表示输出参数，INOUT 表示既可以输入也可以输出。

1. IN 参数

需求：输入数据，返回结果

```
mysql> DELIMITER //
CREATE PROCEDURE in_parameter(IN p_in int)
BEGIN
```

```
select p_in
END
//
DELIMITER
```

调用

```
mysql> call in_parameter(8)\G
*************************** 1. row ***************************
p_in: 8
```

2. OUT 参数

需求：输出当前数据库的用户

```
mysql > DELIMITER //
CREATE PROCEDURE pro_count(OUT out_cnt int)
BEGIN
select count(* ) into out_cnt from mysql. user
END
//
DELIMITER
```

调用

```
mysql> call pro_count(@s)
mysql> select @s\G
*************************** 1. row ***************************
@s: 6
```

3. INOUT 参数

```
mysql > DELIMITER //
CREATE PROCEDURE inout_parameter(INOUT p_inout int)
BEGIN
select p_inout
set p_inout= 2
select p_inout
END
//
DELIMITER;
```

调用

```
mysql> set @p_inout= 1
```

```
mysql> call inout_parameter(@p_inout)\G
*************************** 1. row ***************************
p_inout: 2
1 row in set (0. 00 sec)
```

4.8.3 变量的使用

变量可以在子程序中声明并使用,这些变量的作用范围是 BEGIN...IN 程序中。

(1)定义变量

在存储过程中使用 DECLARE 语句定义变量,语法格式如下:

DECLARE var_name[,varname]... date_type [DEFAULT value]

#总行数
DECLARE cnt INT DEFAULT 0
#循环变量 i
DECLARE i INT DEFAULT 0

(2)为变量赋值

MySQL 中使用 SET 语句为变量赋值,语法格式如下:

SET var_name = expr [,var_name = expr]...

MySQL 中哈可以通过 SELECT...INTO... 为一个或多个变量赋值,语法如下:

SELECT col_name[,...] INTO var_name[,...] table_expr

例如:select count(*) into out_cnt from mysql. user

4.8.4 游标的使用

查询语句可能返回多条记录,如果数据量非常大,需要在存储过程和存储函数中使用游标来逐条读取查询结果集中的记录。游标必须在声明处理程序之前被声明,并且变量和条件还必须在声明游标或处理程序之前被声明,游标只能在存储过程和函数中使用。

(1)声明游标

MySQL 中使用 DECLARE 关键字来声明光标,其语法的基本形式如下:

DECLARE cursor_name CURSOR FOR select_statement

cursor_name 参数表示光标的名称

select_statement 参数表示 SELECT 语句的内容,返回一个用于创建光标的结果集

(2)打开游标

打开光标的语法如下:

OPEN cursor_name{游标名称}

(3)使用游标

使用游标的语法如下:

FETCH cursor_name INTO var_name [, var_name]...{参数名称}

cursor_name 参数表示光标的名称

var_name 参数表示将光标中的 SELECT 语句查询出来的信息存入该参数中,var_name 必须在声明光标之前就定义好。

(4)关闭游标

关闭游标的语句如下:

CLOSE cursor_name{游标名称}

4.8.5 流程控制

流程控制语句用来根据条件控制语句的执行。MySQL 中用来构造流程控制的语句有:IF 语句、CASE 语句、LOOP 语句、LEAVE 语句、ITERATE 语句和 WHILE 语句。

(1)IF 语句:IF 语句用来进行条件判断,根据是否满足条件(可包含多个条件),来执行不同的语句,是流程控制中最常用的判断语句。

下面是一个使用 IF 语句的示例,代码如下:

```
IF age> 25 THEN SET @count1= @count1+ 1
    ELSEIF age= 25 THEN @count2= @count2+ 1
    ELSE @count3= @count3+ 1
END IF
```

(2)CASE 语句:CASE 语句也是用来进行条件判断的,它提供了多个条件进行选择,可以实现比 IF 语句更复杂的判断。

需求:编写一个存储过程,按照输入的参数,若为 1 则插入 1,若为其他值,则插入此值的 2 倍。

```
create table t (id int)
CREATE PROCEDURE proc1 (IN parameter int)
begin
declare var int
set var =  parameter
case var
when 1 then
insert into t values(var)
else
```

```
insert into t values(2* var)
end case
end
```

mysql> set @p_int= 0
mysql> call proc1(@p_int)
mysql> select * from t

（3）LOOP 语句：LOOP 语句可以使某些特定的语句重复执行，与 IF 和 CASE 语句相比，LOOP 只实现了一个简单的循环，并不进行条件判断。

使用 LOOP 语句进行循环操作，代码如下：

```
Add_flag:LOOP
    SET @count= @count+ 1
END LOOP Add_flag
```

注意：LOOP 语句本身没有停止循环的语句，必须使用 LEAVE 语句等才能停止循环，跳出循环过程。

（4）REPEAT 语句：REPEAT 语句是有条件控制的循环语句，每次语句执行完毕后，会对条件表达式进行判断，如果表达式返回值为 TRUE，则循环结束，否则重复执行循环中的语句。

（5）while 语句：WHILE 语句也是有条件控制的循环语句。WHILE 语句和 REPEAT 语句不同的是，WHILE 语句是当满足条件时，执行循环内的语句，否则退出循环。

例：编写一个存储过程，循环插入 1－100。

```
create PROCEDURE proc2()
begin
declare var int
set var = 1
while var < = 100 do
insert into t2 values(var)
set var = var+ 1
end while
end
```

mysql> call proc2()

（6）ITERATE 语句：ITERATE 是"再次循环"的意思，用来跳出本次循环，直接进入下一次循环。

注意：LEAVE 语句和 ITERATE 语句都用来跳出循环语句，但两者的功能是不一样的。LEAVE 语句是跳出整个循环，然后执行循环后面的程序。而 ITERATE

语句是跳出本次循环,然后进入下一次循环。

4.8.6 存储过程其他操作

1. 查看存储过程

mysql> show create PROCEDURE myproc
mysql> select a. ROUTINE_TYPE,a. SPECIFIC_NAME from information_schema. ROUTINES a
where a. ROUTINE_SCHEMA = 'jeames'

2. 修改存储过程

mysql> ALTER PROCEDURE showstuscore MODIFIES SQL DATA SQL SECURITY INVOKER;

①以上修改存储过程后,访问数据的权限已经变成了 MODIFIES SQL DATA,安全类型也变成了 INVOKE

②ALTER PROCEDURE 语句用于修改存储过程的某些特征,如果要修改存储过程的内容,应先删除原存储过程后重建

3. 删除存储过程

mysql> drop PROCEDURE pro_count_mysql

4. 调用存储过程

```
#无参数
call proc2()
#有参数
mysql> set @p_int= 0;
mysql> call proc1(@p_int)
```

4.8.7 存储函数

在 MySQL 中,存储函数的使用方法与 MySQL 内部函数的使用方法是一样的。换言之,用户自己定义的存储函数与 MySQL 内部函数的性质相同。区别在于,存储函数是用户自己定义的,而内部函数是 MySQL 开发者定义的。

创建存储函数,需要使用 CREATE FUNCTION 语句,函数体必须包含一个 RETURN value 语句,基本语法格式如下:

CREATE FUNCTION func_name([func_parameter])
RETURNS type[characteristic ...]
routine_body

CREATE FUNCTION 为用来创建存储函数的关键字；func_name 表示存储函数的名称；func_parameter 为存储过程的参数列表，FUNCTION 中总是默认为 IN 参数，参数列表形式如下：

[IN|OUT|INOUT] param_name type

案例：编写一个函数实现在某一个字符串后面循环拼接 n 个其他字符串，例如：假设该函数名为 fn_addStr(str1 varchar(20),str2 varchar(20),int)，则调用后 Selecet fn_addStr('Hello','World',4)

显示：HelloWorldWorldWorldWorld

```
create function fun_addStr(str1 varchar(100), str2 varchar(10), num int) returns varchar(200)
begin
declare i int default 1
declare result varchar(200) default''
set result= str1
myloop:loop
set i= i+ 1
set result= concat(result,str2)
if i> num
then
leave myloop
end if
end loop myloop
return result;/end

## 调用函数
select fun_addStr('Hello','World',8)
```

4.9 触发器

触发器是由事件来触发某个操作，这些事件包括 INSERT、UPDATE 和 DELETE 语句。如果定义了触发程序，当数据库执行这些语句的时候就会激发触发器执行相应的操作，触发程序是与表有关的命名数据库库对象，当表上出现特定事件时，将激活该对象，可以实现更负责，更精细的管理控制。

4.9.1 创建触发器

可以使用 CREATE TRIGGER 语句创建触发器，语法格式如下：

```
CREATE <触发器名> < BEFORE | AFTER >
<INSERT | UPDATE | DELETE >
ON <表名> FOR EACH Row<触发器主体>
```

语法说明如下。

(1)创建触发器的三要素。

①监视地点:表

②监视的事件:insert\update\delete

③触发的时间:after、before

(2)FOR EACH ROW 一般是指行级触发,对于受触发事件影响的每一行都要激活触发器的动作。

(3)每个表的每个事件每次只允许有一个触发器,单一触发器不能与多个事件或多个表关联。

案例:一张表记录了姓名及收入,当插入数据的时候,一张统计总人数及收入。

```
##建2个表
create table d1 (id int,name varchar(10),salary int);
create table d2 (user_total int,salary_total int);

##插入数据
insert into d1 values(1,'xiaoming',3000),(2,'xiaoli',4000),(3,'xiaohei'3500),(4,'xiaohu',1000)

##创建触发器
--插入数据时
create trigger d1_ai
after insert on d1 for each row
update d2 set user_total = user_total + 1,salary_total = salary_total + new.salary;

--删除数据时
create trigger d1_ad
after delete on d1 for each row
update d2 set user_total = user_total - 1,salary_total = salary_total - old.salary;

--更新操作时
create trigger d1_au
after update on d1 for each row
update d2 set salary_total = salary_total - old.salary + new.salary
```

4.9.2 查看触发器

查看触发器是指查看数据库中已存在的触发器的定义、状态和语法信息等。可以通过命令来查看已经创建的触发器,触发器的数据字典如图4-8所示。

通过 SHOW TRIGGERS 查看触发器的语句如下:

①通过 SHOW TRIGGERS 查看触发器

```
mysql> show triggers\G
```

②在 triggers 表中查看触发器

MySQL 中所有触发器的定义都存在 INFORMATION_SCHEMA 数据库 TRIGGERS 数据字典中,如图4-9所示。

```
mysql> select *  from information_schema.'TRIGGERS'
```

ACTION_ORDER	ACTION_TIMING	TRIGGER_SCHEMA	EVENT_MANIPUL	ACTION_STATEMENT
1	BEFORE	sys	INSERT	BEGIN　IF @sys.ignore_sys_config_triggers != true AND NEW.set_by IS NULL THEN
1	BEFORE	sys	UPDATE	BEGIN　IF @sys.ignore_sys_config_triggers != true AND NEW.set_by IS NULL THEN
1	AFTER	jeames	INSERT	update d2 set user_total=user_total+1,salary_total=salary_total+new.salary
1	AFTER	jeames	DELETE	update d2 set user_total=user_total-1,salary_total=salary_total-old.salary
1	AFTER	jeames	UPDATE	update d2 set salary_total=salary_total-old.salary+new.salary

图4-9　触发器数据字典 TRIGGERS

4.9.3 修改及删除触发器

修改触发器可以删除原触发器,再以相同的名称创建新的触发器,使用 DROP 语句将触发器从数据库中删除,语法格式如下:

```
DROP TRIGGER [ IF EXISTS ][数据库名]<触发器名>
```

删除一个表的同时,也会自动删除该表上的触发器。另外,触发器不能更新或覆盖,如要修改一个触发器,必须先删除它,再重新创建。

4.10 事件(event)

事件和触发器类似,都是在某些事情发生的时候启动,事件是根据调度事件来启动的,事件取代了原先只能由操作系统计划任务来执行的工作,而且 MySQL 的事件调度器可以精确到每秒钟执行一个任务,而操作系统的计划任务(如 Linux 下的 crontab 或 Windows 下的任务计划)只能精确到每分钟执行一次。

4.10.1 事件配置

如果事件调度程序状态未设置为禁用,则可以在 ON 和 OFF 之间切换 Event_Scheduler(使用 set)。设置此变量时,也可以使用0表示关闭,使用1表示打开。

```
SHOW VARIABLES LIKE '%event_sche%'
```

因此,可以在 MySQL 客户端中使用以下 4 条语句中的任意一条来打开事件调度程序:

```
mysql> SET GLOBAL event_scheduler = ON
mysql> SET @@GLOBAL.event_scheduler = ON
mysql> SET GLOBAL event_scheduler = 1
mysql> SET @@GLOBAL.event_scheduler = 1
```

类似地,这 4 条语句中的任何一条都可以用于关闭事件调度程序:

```
mysql> SET GLOBAL event_scheduler = OFF
mysql> SET @@GLOBAL.event_scheduler = OFF
mysql> SET GLOBAL event_scheduler = 0
mysql> SET @@GLOBAL.event_scheduler = 0
```

要禁用事件计划程序,可使用以下两种方法:

(1)作为启动服务器时的命令行选项:

```
-- event-scheduler= DISABLED
```

(2)在服务器配置文件(my.cnf 或 Windows 系统上的 my.ini)中

```
event_scheduler= DISABLED
```

4.10.2 事件创建

语句创建并调度新事件时,除非启用事件计划程序,否则事件不会运行,有效创建事件语句的最低要求如下:

(1)关键字 CREATE EVENT 要加上一个事件名称,该名称应为唯一标识数据库架构中的事件。

(2)用 ON SCHEDULE 子句确定事件执行的时间和频率。

(3)DO 子句要包含由事件执行的 SQL 语句。

```
CREATE
    [DEFINER = user]
    EVENT
    [IF NOT EXISTS]
    event_name
    ON SCHEDULE schedule
    [ON COMPLETION [NOT] PRESERVE]
    [ENABLE | DISABLE | DISABLE ON SLAVE]
```

```
    [COMMENT 'string']
    DO event_body;

schedule: {
    AT timestamp [+ INTERVAL interval] ...
  | EVERY interval
    [STARTS timestamp [+ INTERVAL interval] ... ]
    [ENDS timestamp [+ INTERVAL interval] ... ]
}

interval:
    quantity {YEAR | QUARTER | MONTH | DAY | HOUR | MINUTE |
              WEEK | SECOND | YEAR_MONTH | DAY_HOUR | DAY_MINUTE |
              DAY_SECOND | HOUR_MINUTE | HOUR_SECOND | MINUTE_SECOND}
```

案例：以下为简单创建事件语句的示例。

```
CREATE EVENT myevent
    ON SCHEDULE AT CURRENT_TIMESTAMP + INTERVAL 1 HOUR
    DO
        UPDATE myschema.mytable SET mycol = mycol + 1
```

说明：

DEFINER 子句指定在事件执行时检查访问权限时使用的 MySQL 帐户。

如果 event_name 中未指定任何模式，则假定默认（当前）模式。

ON SCHEDULE 子句确定为事件定义的 event_body 重复的时间、频率和持续时间。

AT 时间戳用于一次性事件。它指定事件仅在时间戳指定的日期和时间执行一次，时间戳必须包括日期和时间，或者必须是解析为日期时间值的表达式，EVERY 表示一定频率。

4.10.3 事件操作

（1）临时关闭事件

```
ALTER EVENT e_test DISABLE
```

（2）开启事件

```
ALTER EVENT e_test ENABLE
```

（3）删除事件（DROP EVENT）

```
DROP EVENT [IF EXISTS] event_name
```

4.11 Prometheus 监控

Prometheus(由 go 语言(golang)开发)是一套开源的监控 & 报警 & 时间序列数据库的组合。适合监控 docker 容器。因为 kubernetes(俗称 k8s)的流行带动了 Prometheus 的发展。

Prometheus 提供了从指标暴露,到指标抓取、存储和可视化,以及最后的监控告警等组件,如图 4-10 所示。

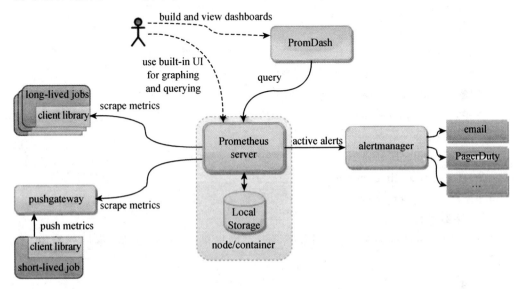

图 4-10 Prometheus 架构组成

从图 4-9 可发现,Prometheus 整个生态圈组成主要包括 Prometheus server、Exporter、Pushgateway、Alertmanager、Grafana、Web ui 界面,其中 Prometheus server 由三个部分组成,分别为 Retrieval、Storage、PromQL。

Retrieval 负责在活跃的 target 主机上抓取监控指标数据。

Storage 存储主要是把采集到的数据存储到磁盘中。

PromQL 是 Prometheus 提供的查询语言模块。

官网:https://prometheus.io/。

Prometheus 原理总结如下:

(1)Prometheus server 可定期从活跃的(up)目标主机上(target)拉取监控指标数据,目标主机的监控数据可通过配置静态 job 或服务发现的方式被 prometheus server 采集到,这种方式默认的 pull 方式可拉取指标;也可通过 Pushgateway 把采集的数据上报到 Prometheus server 中;还可通过一些组件自带的 exporter 采集相应组件的数据。

(2)Prometheus server 把采集到的监控指标数据保存到本地磁盘或者数据库。

(3)Prometheus 采集的监控指标数据按时间序列存储,通过配置报警规则,把触发的报警发送到 alertmanager。

(4)Alertmanager 通过配置报警接收方,发送报警到邮件,微信或者钉钉等。

(5)Prometheus 自带的 Web ui 界面提供 PromQL 查询语言,可查询监控数据。

(6)Grafana 可接入 prometheus 数据源,把监控数据以图形化形式展示出。

4.11.1 安装 Prometheus

1. 端口程序

9090   prometheus

3000   grafana

9093   alter_manager

9100   node_exporter

9104   mysqld_exporter

9121   redis_exporter

9161   oracledb_exporter

9216   mongodb_exporter

9187   postgres_exporter

2. 下载地址

从官网下载相应版本,安装到服务器上官网提供的是二进制版,解压就能用,不需要编译。

官网地址:

https://prometheus.io/download/.

监控组件下载地址:

https://prometheus.io/docs/instrumenting/exporters/.

可通过以下方式解压启动:

tar – zxvf prometheus – 2.45.3.linux – amd64.tar.gz – C /usr/local/

ln – s /usr/local/prometheus – 2.45.3.linux – amd64 /usr/local/prometheus

ln – s /usr/local/prometheus/prometheus /usr/local/bin/prometheus

prometheus —— config.file = /usr/local/prometheus/prometheus.yml \

—— storage.tsdb.path = /usr/local/prometheus/data/ \

—— web.enable – lifecycle \

—— storage.tsdb.retention.time = 15d &

#lsof – i:9090   查看端口

#ps – ef|grep prometheus 查看进程

第4章 DBMS 基本管理

端口为9090,访问网址根据实际 IP 更改

http://192.168.3.10:9090.

详细的参数解释如下：

——web.enable-lifecycle

加上此参数可远程加载配置文件,无需重启 prometheus。

调用指令是 curl-X POST http://ip:9090/-/reload

——storage.tsdb.retention.time

数据默认保存时间为 15 天,启动时加上此参数可控制数据保存时间。

-config.file=/etc/prometheus.yml 指定配置文件。

-web.read-timeout=5m 请求链接的最大等待时间,防止太多的空闲链接占用资源。

-web.max-connections=512 针对 prometheus,获取数据源的时候,建立的网络链接数,应做一个最大数字的限制,以防止链接数过多造成资源消耗过大。

-storage.tsdb.retention=15d 重要参数,prometheus 开始采集监控数据后,会存在内存和硬盘中;对于保存期限的设置,时间过长,硬盘和内存都吃不消;时间太短,要查历史数据就没了。15 天最为合适。

-storage.tsdb.path="/prometheus/data" 存储数据路径,不可随便定义。

-query.max-concurrency=20 用户查询最大并发数。

-query.timeout=2m 慢查询强制终止。

通过访问 http://192.168.3.10:9090/targets 此地址,可以查看页面中的 Targets 信息。示例中使用默认配置文件,仅仅对 Prometheus 本机进行监控如图 4-11 所示。

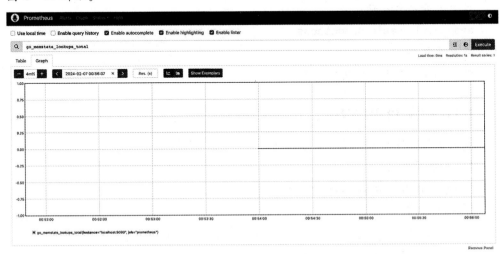

图 4-11 Prometheus 监控界面

注意：配置文件 prometheus.yml 不能加双引号，否则启动报错，找不到文件或目录。本次启动用户是 root。生产中最好新建一个用户用于启动，需要设置配置文件及数据文件权限数据目录。另外，最好单独配置数据硬盘，例如可使用 LVM 硬盘格式配置。

4.11.3 开机自启动

```
＃cat /usr/lib/systemd/system/prometheus.service
[Unit]
Description=prometheus
[Service]
ExecStart=/usr/local/prometheus/prometheus --config.file=/usr/local/prometheus/prometheus.yml --storage.tsdb.path=/usr/local/prometheus/data/ --web.enable-lifecycle --storage.tsdb.retention.time=15d
ExecReload=/bin/kill -HUP $MAINPID
KillMode=process
Restart=on-failure
[Install]
WantedBy=multi-user.target
＃通知 systemd 重新加载配置文件
systemctl daemon-reload
＃启动
systemctl start prometheus
＃设置开机自启动
systemctl enable prometheus
＃查看状态
systemctl status prometheus
```

4.11.4 监控 MySQL

安装在要监控的 MySQL 服务器即可。

(1) 下载安装包

https://prometheus.io/download/

(2) 解压安装

```
tar -zxvf mysqld_exporter-0.15.1.linux-amd64.tar.gz
mv ./mysqld_exporter-0.15.1.linux-amd64/mysqld_exporter /usr/local/bin/
```

(3)添加数据源

①创建监控用户。

mysql> create user mysql_exporter@'%' identified with mysql_native_password by 'root';

mysql> grant process, replication client, select on *.* to 'mysql_exporter'@'%';

mysql> flush privileges;

②配置 MySQL 密码。

```
mkdir -p /etc/mysql_exporter
cd /etc/mysql_exporter
cat > mysql3306_192168312.cnf << "EOF"
[client]
host = 192.168.3.12
user = mysql_exporter
password = root
port = 3306
EOF
```

(4)添加监控目标

```
vi /usr/local/prometheus/prometheus.yml
- job_name: 'MySQL'
  static_configs:
  - targets: ['192.168.3.12:9104']
    labels:
      instance: mysql_Prometheus
systemctl restart prometheus
systemctl status prometheus
```

在 Grafana 官网中找到 mysql 的监控模板 Mysql Overview,ID 为 7362,使用 ID 将其导入到 Grafana 中。

Grafana 导入 mysql_exporter 的模板,监控界面如图 4-12 所示。

主从监控模板 ID:7371

MySQL 状态监控模板 ID:7362

缓冲池状态模板 ID:7365

获取数据源,端口 9104:

curl http://192.168.3.10:9104/metrics

图 4-12　grafana 结合 prometheus 监控界面

4.12　Zabbix 监控

4.12.1　Zabbix 简介

Zabbix 是一个基于 WEB 界面的提供分布式系统监视以及网络监视功能企业级的开源解决方案。它是一个企业级的高度集成开源监控软件，可以用来监控设备、服务器、数据库等可用性和性能，保证服务器系统及数据库的安全运营。并提供灵活的通知机制，以让系统管理员快速定位/解决存在的各种问题。

Zabbix 监控的优点：

(1)自动发现服务器和网络设备。

(2)底层自动发现(如自动发现多实例 Mysql、Tomcat 进程等)。

(3)分布式的监控体系和集中式的 web 管理。

(4)支持主动监控和被动监控模式。

(5)支持多种操作系统 linux，Solaris，HP－UX，AIX，FreeBSD，OpenBSD 等。

(6)高效的 Agent 支持 linux，Solaris，HP－UX，AIX，FreeBSD，OpenBSD，windows NT4.0，window2000 等。

(7)无 Agent 监控等多种监控方法(如用 SNMP 协议监控路由或交换机、IPMI 检测硬件温度等)。

(8)安全的用户认证模式。

(9)灵活的用户权限设置。

(10)基于 web 的管理方法，支持自由的定义事件和邮件发送。

(11)高水平的业务视图监控资源,支持日志审计。

Zabbix 官网如图 4-13 所示。

图 4-13　Zabbix 官网界面

4.12.2　Zabbix 构成

zabbix 主要由以下 5 个组件构成:

1. Server(服务端)

zabbix server 是 zabbix 的核心组件,server 内部存储了所有的配置信息、统计信息和操作信息。zabbix agent 会向 zabbix server 报告可用性、完整性及其他统计信息。

2. web 页面(网站服务)

web 页面也是 zabbix 的一部分,通常和 zabbix server 位于一台物理设备上,但是在特殊情况下也可以分开配置。web 页面主要提供了直观的监控信息,以方便运维人员监控管理。

3. 数据库(存储数据)

zabbix 数据库内存储了配置信息、统计信息等 zabbix 的相关内容。

4. proxy(HA 功能)

zabbix proxy 可以根据具体生产环境采用或者放弃。如果使用了 zabbix proxy,则其会替代 zabbix server 采集数据信息,可以很好分担 zabbix server 的负载。zabbix proxy 通常运用于架构过大、zabbix server 负载过重,或者是企业设备跨机房、跨网段、zabbix server 无法与 zabbix agent 直接通信的场景。

5. Agent(客户端)

zabbix agent 通常部署在被监控目标上,用于主动监控本地资源和应用程序,并将监控的数据发送给 zabbix server。

4.12.3 Zabbix 部署

本次部署采用在线安装方式,操作系统为 centos8.2,MySQL 版本为 8.0 来部署 zabbix6.2,其官方环境如 4-14 所示。

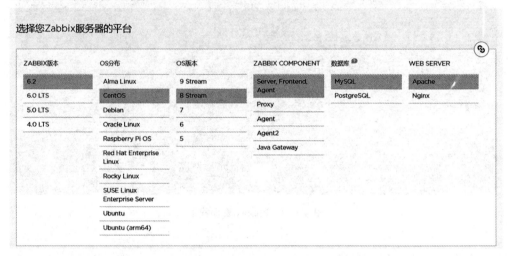

图 4-14 Zabbix 官方环境

1. 配置 zabbix 源仓库

Zabbix 源仓库安装前,需要配置 yum,自 2022 年 1 月 31 日起,CentOS 团队从官方镜像中移除了 CentOS 8 的所有包,如需继续运行旧 CentOS 8,可以在 /etc/yum.repos 中更新 repos.d,使用 vault.centos.org 代替 mirror.centos.org。

可使用以下方法替换:

[root@jemzabbix6 ~]# cd /etc/yum.repos.d/

[root@jemzabbix6 ~]# sed -i 's/mirrorlist/#mirrorlist/g' /etc/yum.repos.d/CentOS-*

[root@jemzabbix6 ~]# sed -i 's|#baseurl=http://mirror.centos.org|baseurl=http://vault.centos.org|g' \
/etc/yum.repos.d/CentOS-*

[root@jemzabbix6 ~]# rpm -Uvh \
https://repo.zabbix.com/zabbix/6.2/rhel/8/x86_64/zabbix-release-6.2-1.el8.noarch.rpm

2. 切换 PHP 的 DNF 版本

CentOS8 给我们带来了一些新的特性,可以完全用 dnf 取代 yum 来进行包管理。dnf module 在软件安装上更方便,可以通过 dnf module install 在安装软件时指定安装的版本,默认 dnf install 安装时,优先安装软件仓库中的最新版本。

[root@jemzabbix6 ~]# dnf module list
[root@jemzabbix6 ~]# dnf module enable php:7.4

3. 安装 Zabbix 组件

安装 Zabbix server,Web 前端,agent。

[root@jemzabbix6 ~]# dnf install zabbix-server-mysql \
zabbix-web-mysql \
zabbix-apache-conf \
zabbix-sql-scripts \
zabbix-selinux-policy \
zabbix-agent

查看已经安装的 zabbix 组件。

[root@jemzabbix6 ~]# rpm -qa | grep zabbix
zabbix-web-deps-6.2.2-release1.el8.noarch
zabbix-server-mysql-6.2.2-release1.el8.x86_64
zabbix-agent-6.2.2-release1.el8.x86_64
zabbix-release-6.2-1.el8.noarch
zabbix-web-6.2.2-release1.el8.noarch
zabbix-web-mysql-6.2.2-release1.el8.noarch
zabbix-apache-conf-6.2.2-release1.el8.noarch
zabbix-selinux-policy-6.2.2-release1.el8.x86_64
zabbix-sql-scripts-6.2.2-release1.el8.noarch

4. 创建初始数据库

MySQL 数据库可用来给 Zabbix Server 提供数据存储,本次采用 yum 在线 rpm 安装 MySQL8.0。

(1)安装 MySQL8.0

①repo 下载

rpm -Uvh https://repo.mysql.com//mysql80-community-release-el8.rpm

②查询 yum 里的 MySQL 版本

[root@jemzabbix6 ~]# yum repolist all | grep mysql

③配置安装的 MySQL 的版本

[root@jemzabbix6 ~]# yum -y install yum-utils
[root@jemzabbix6 ~]# yum-config-manager --enable mysql80-community

④查询安装的 MySQL 的版本

[root@jemzabbix6 ~]# yum repolist enabled | grep mysql

mysql-connectors-community   MySQL Connectors Community
mysql-tools-community   MySQL Tools Community
mysql80-community   MySQL 8.0 Community Server

⑤安装 MySQL

先执行：yum module disable mysql

再执行：yum -y install mysql-community-server

yum install -y mysql-community-server

⑥初始化 MySQL

[root@jemzabbix6 ~]# systemctl start mysqld

⑦查看 MySQL 状态

[root@jemzabbix6 ~]# systemctl status mysqld

⑧查看临时密码

[root@jemzabbix6 ~]# grep 'temporary password' /var/log/mysqld.log

⑨登录 MySQL 后修改密码

[root@jemzabbix6 ~]# mysql -uroot -p

mysql> ALTER USER root@'localhost' IDENTIFIED BY 'Jeames@007';

mysql> flush privileges;

(2)创建用户配置参数

mysql> create database zabbix character set utf8mb4 collate utf8mb4_bin;

mysql> create user zabbix@localhost identified by 'Jeames@007';

mysql> grant all privileges on zabbix.* to zabbix@localhost;

mysql> set global log_bin_trust_function_creators = 1;

mysql> quit;

简单介绍一下 log_bin_trust_function_creators 参数：当二进制日志启用后，变量就会启用。它可以控制是否信任存储函数创建者，但不会创建写入二进制日志引起不安全事件的存储函数。如果设置为 0（默认值），用户不得创建或修改存储函数。这里我们修改为 1。

(3)导入初始架构和数据，系统将提示您输入新创建的密码

[root@jemzabbix6 ~]# zcat /usr/share/doc/zabbix-sql-scripts/mysql/server.sql.gz | \
mysql --default-character-set=utf8mb4 -uzabbix -p zabbix

(4) log_bin_trust_function_creators 参数关闭

mysql> set global log_bin_trust_function_creators = 0；
mysql> select @@log_bin_trust_function_creators；

(5) 为 Zabbix server 配置数据库

编辑配置文件 /etc/zabbix/zabbix_server.conf
#这个要输入的密码就是在 MySQL 中创建 zabbix 数据配置的密码
DBPassword = Jeames@007

5. 启动 Zabbix 相关进程

启动 Zabbix server 和 agent 进程，并为它们设置开机自启。

[root@jemzabbix6 ~]# systemctl restart zabbix-server zabbix-agent httpd php-fpm

[root@jemzabbix6 ~]# systemctl enable zabbix-server zabbix-agent httpd php-fpm

查看服务状态，如图 4-15 所示。

[root@jemzabbix6 ~]# systemctl status zabbix-server zabbix-agent httpd php-fpm | grep Active -B 3

```
[root@jemzabbix6 ~]# systemctl status zabbix-server zabbix-agent httpd php-fpm | grep Active -B 3
● zabbix-server.service - Zabbix Server
   Loaded: loaded (/usr/lib/systemd/system/zabbix-server.service; enabled; vendor preset: disabled)
--
   Active: active (running) since Sat 2022-09-10 11:42:10 UTC; 1min 30s ago

● zabbix-agent.service - Zabbix Agent
   Loaded: loaded (/usr/lib/systemd/system/zabbix-agent.service; enabled; vendor preset: disabled)
   Active: active (running) since Sat 2022-09-10 11:42:09 UTC; 1min 30s ago

   Loaded: loaded (/usr/lib/systemd/system/httpd.service; enabled; vendor preset: disabled)
   Drop-In: /usr/lib/systemd/system/httpd.service.d
            └─php-fpm.conf
   Active: active (running) since Sat 2022-09-10 11:42:10 UTC; 1min 29s ago

● php-fpm.service - The PHP FastCGI Process Manager
   Loaded: loaded (/usr/lib/systemd/system/php-fpm.service; enabled; vendor preset: disabled)
   Active: active (running) since Sat 2022-09-10 11:42:10 UTC; 1min 29s ago
```

图 4-15 Zabbix 相关进程

后期 zabbix 会涉及语言的设置，配置 Linux 环境语言方法如下：
安装中文语言包，命令 yum install glibc-langpack-zh
安装英文语言包，命令 dnf install langpacks-en glibc-all-langpacks-y
安装一个 glic-common 底层库对语言进行重新识别，[root@jemzabbix6 ~]# dnf install glibc-common

6. Zabbix 配置

IP 加端口号，端口号默认为 80，下一步执行即可。

http://192.168.1.54:280/zabbix

Zabbix 主界面和配置如图 4-16、图 4-17 所示。默认用户名密码：Admin/zabbix。

接下来配置 zabbix 主机名称、默认时区（选上海）、默认主题如图 4-18、如图 4-19 所示。

Configuration file "/etc/zabbix/web/zabbix.conf.php" created. 后期可修改端口号等。

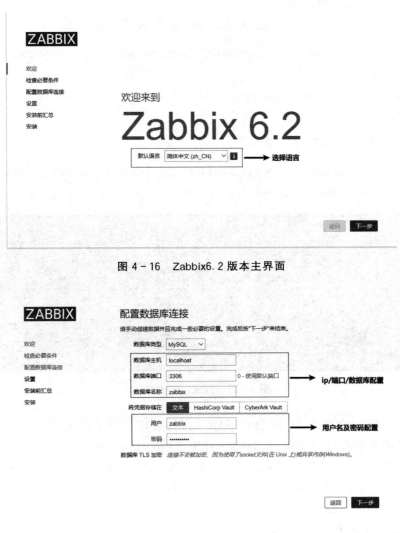

图 4-16　Zabbix6.2 版本主界面

图 4-17　Zabbix 数据库配置界面

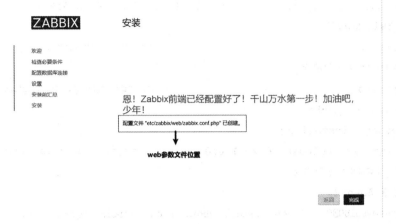

图 4-18 Zabbix 配置成功界面

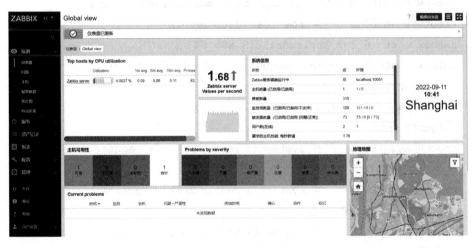

图 4-19 Zabbix 监控主界面

7. Zabbix 监控 MySQL

本次监控是基于 MySQL8.0 版本,在 Linux 环境下 MySQL 服务器环境下做以下操作。

(1)安装 Agent2

① ## 配置 zabbix 源仓库

[root@mysql8030 ~]# rpm -Uvh \

https://repo.zabbix.com/zabbix/6.2/rhel/8/x86_64/zabbix-release-6.2-1.el8.noarch.rpm

Retrieving https://repo.zabbix.com/zabbix/6.2/rhel/8/x86_64/zabbix-release-6.2-1.el8.noarch.rpm

```
warning: /var/tmp/rpm-tmp.tRFyj2: Header V4 RSA/SHA512 Signature, key ID a14fe591: NOKEY
Verifying...                              ################################## [100%]
Preparing...                              ################################## [100%]
Updating / installing...
   1:zabbix-release-6.2-1.el8            ################################## [100%]
```

②安装 Agent2

```
[root@mysql8030 ~]# dnf install zabbix-agent2 zabbix-agent2-plugin-mongodb
[root@mysql8030 zabbix]# systemctl restart zabbix-agent2
[root@mysql8030 zabbix]# systemctl enable zabbix-agent2
[root@mysql8030 ~]# rpm -qa | grep zabbix
zabbix-get-6.2.2-release1.el8.x86_64
zabbix-agent2-plugin-mongodb-1.0.0-1.el8.x86_64
zabbix-release-6.2-1.el8.noarch
zabbix-agent2-6.2.2-release1.el8.x86_64
```

③Agent2 进程

```
[root@mysql8030 ~]# systemctl restart zabbix-agent2
##设置开机自启动
[root@mysql8030 ~]# systemctl enable zabbix-agent2
##查看 Agent 进程状态
[root@mysql8030 ~]# systemctl status zabbix-agent2
```

④配置 zabbix-agent.conf 文件

```
[root@jemzabbix6 ~]# vi /etc/zabbix/zabbix_agent2.conf
Server = 172.18.0.3
ServerActive = 172.18.0.4
Hostname = mysql8030
Timeout = 10
```

参数说明：

server 改成 zabbix 服务端的 IP 地址。

serveractive 改成 zabbix 客户端的 IP 地址。

hostname 改成本机的主机名，或本机的 IP 地址。

⑤重启进程生效

```
[root@mysql8030 ~]# systemctl restart zabbix-agent2
```

（2）Zabbix Server 添加主机

配置—>主机—>创建主机，Zabbix6 自带 MySQL 模板，无须再使用第三方，也不用自己写监控脚本，而且模板 zabbix‑agent2 比 zabbix‑agent，功能更强大，应用也更简单，如图 4‑20 所示。

登录 Zabbix Server WEB，【配置】→【模版】—【MySQL by Zabbix agent 2】，修改此【宏】，如图 4‑21～图 4‑22 所示。

图 4‑20　Zabbix 添加 MySQL 主机

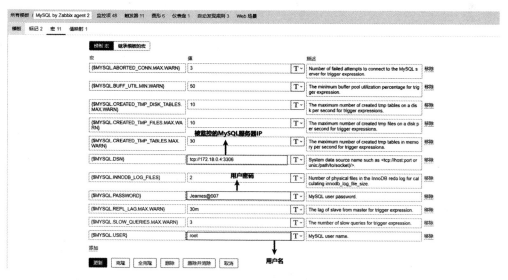

图 4‑21　MySQL 监控模板参数界面

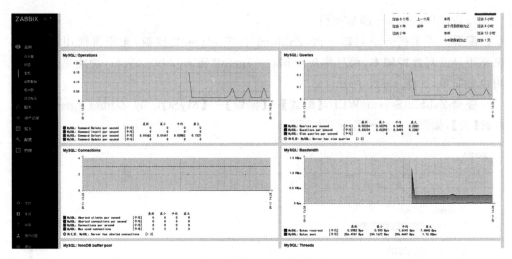

图 4-22　MySQL 监控 MySQL 各项性能

Part 3　体系架构篇

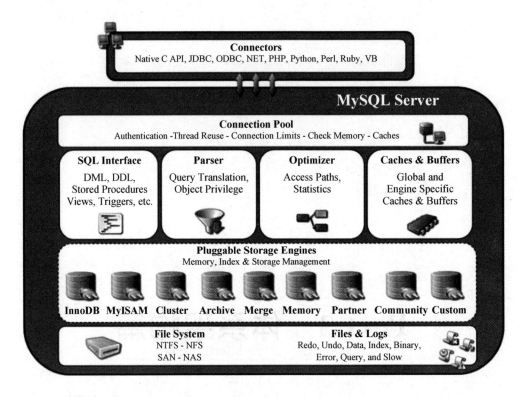

1. 连接者（Connectors）

连接者指的是不同语言中与 SQL 的交互者,目前流行的语言都支持 MySQL 客户端连接。

2. 连接池（Connection Pool）

连接池管理缓冲用户连接、线程处理等需要缓存的需求。在这里也会进行用户账号、密码和库表权限验证。

3. SQL 接口（SQL Interface）

SQL 接口接收用户执行的 SQL 语句,并返回查询的结果。

4. 查询解析器（Parser）

SQL 语句被传递到解析器时会进行验证和解析（解析成 MySQL 认识的语法,查询什么表、什么字段）,解析器是由 Lex 和 YACC 实现的一个很长的脚本。其主要功能是将 SQL 语句分解成数据结构,并将这个结构传递到后续步骤中,后续 SQL 语句的传递和处理就是基于这个结构的。如果在分解构成中遇到错误,则说明该 SQL 语句可能有语法错误或者不合理。

Lex 工具是一种词法分析程序生成器,它可以根据词法规则说明书的要求来生成单词识别程序,由该程序识别出输入文本中的各个单词。如同 Lex 一样,一个 YACC 程序也用双百分号分为三段,它们是声明、语法规则和 C 代码,Yacc 用来创建一个编译器。

5. 查询优化器（Optimizer）

优化器的主要作用是将 SQL 经过词法解析/语法解析后得到的语法树，通过 MySQL 的数据字典和统计信息的内容，经过一系列运算，最终得出一个执行计划，包括选择使用哪个索引等。

6. 缓存或缓冲（Caches & Buffers）

缓存或缓冲主要包含 QC(QueryCache) 以及表缓存、权限缓存等。对于 QC，以往主要用于 MyISAM 存储引擎，目前 MySQL 8.0 中已放弃，对于现在非常流行的 InnoDB 存储引擎来讲，QC 已无任何意义，因为 InnoDB 存储引擎有自己的且非常完善的缓存功能。除 QC 之外（记录缓存、key 缓存，可使用参数单独关闭），该缓存机制还包括表缓存和权限缓存等，这些属于 Server 层的功能，其他存储引擎仍需要使用。

7. 插件式存储引擎（Pluggable Storage Engines）

存储引擎是 MySQL 中具体的与文件打交道的子系统，也是 MySQL 最具特色的一个地方。MySQL 的存储引擎是插件式的，它根据 MySQL AB 公司提供的文件访问层的一个抽象接口来定制一种文件访问机制（这种访问机制就叫存储引擎）。目前存储引擎众多，且它们的优势各不相同，现在最常用于 OLTP 场景的是 InnoDB（当然也支持 OLAP 存储引擎，但 MySQL 自身的机制并不擅长 OLAP 场景）。

8. 磁盘物理文件（Files & Logs）

磁盘物理文件包含 MySQL 的各个引擎的数据、索引的文件，以及 redo log、undo log、binary log、error log、query log、slow log 等各种日志文件。

9. 文件系统（File System）

文件系统是对存储设备的空间进行组织和分配，负责文件存储，并对存入的文件进行保护和检索的系统。它负责为用户建立文件，存入、读出、修改、转储文件，控制文件的存取。常见的文件系统包括 XFS、NTFS、EXT4、EXT3、NFS 等，通常数据库服务器使用磁盘建议用 XFS。

第 5 章 物理结构

MySQL 物理结构主要包括以下文件类型:
(1)参数文件:MySQL 实例启动的时候在哪里可以找到数据库文件,并且指定某些初始化参数,这些参数定义了某种内存结构的大小等设置,以及参数类型和定义作用域。
(2)日志文件:记录 MySQL 对某种条件做出响应时候写入的文件,包括错误日志、慢查询日志、全查询日志、二进制日志、中继日志、Redo 日志、Undo 日志。
(3)MySQL 表结构文件:存放 MySQL 表结构定义的文件。
(4)存储引擎文件:记录存储引擎信息的文件。
Socket 文件:当用 Linux 的 MySQL 命令行窗口登录的时候需要的文件。
Pid 文件:MySQL 实例的进程文件。
auto.cnf:MySQL 实例全局唯一的 server-uuid 文件,主要用于主从复制中。
*.pem:MySQ 的证书文件,用于 ssl 认证。

5.1 参数文件

当 MySQL 实例启动时,MySQL 会先去读一个配置参数文件,用来寻找数据库各种文件所在的位置以及指定某些初始化参数,这些参数通常定义了某种内存结构有多大等设置。

MySQL 参数文件的作用和 Oracle 的参数文件极其类似。不同的是,Oracle 实例启动时若找不到参数文件,是不能进行装载(mount)操作的。MySQL 稍微有所不同,MySQL 实例可以不需要参数文件,这时所有的参数值取决于编译 MySQL 时指定的默认值和源代码中指定参数的默认值。

5.1.1 参数文件的位置

默认情况下,MySQL 实例会按照一定的次序去取,在 Linux 环境中,通过命令 mysql --help|grep my.cnf 来寻找即可。

[root@jeames ~]# mysql --help | grep 'my.cnf'
/etc/my.cnf /etc/mysql/my.cnf /usr/local/mysql/etc/my.cnf ~/.my.cnf

有些 DBA 曾问我,在 Windows 环境下 MySQL 的参数在哪里?默认情况下 MySQL 参数就放在数据存放目录下。另外一种方法:在"开始 → 所有程序 → MySQL"下面找到 MySQL 的命令行客户端工具,然后选择该命令行工具,查看"属性"，在"目标"里面也可以看到 MySQL 使用的配置文件位置。

大家可以把数据库参数看成一个键/值对，通过 show variables 查看所有的参数，或通过 like 来过滤参数名。从 MySQL 5.1 版本开始，已可以通过 information_schema 架构下的 GLOBAL_VARIABLES 视图来进行查找。

5.1.2 参数文件类别

MySQL 的参数根据是否可以在数据库运行时修改，可分为 2 类：

动态参数（dynamic）。动态参数表示可以在 MySQL 运行时修改其参数值，并且可立即生效。

静态参数（static）。静态参数则表示在 MySQL 实例运行期间不能修改只能只读的参数。若要修改生效，需要先修改配置文件，再重启 MySQL 实例，若修改静态参数则会报错。

1. 动态改变

关键字 global 和 session 可用来标记该参数基于会话还是整个实例。

有些参数支持会话级动态修改，如 autocommit Mysql> set SESSION 等。

有些参数支持 global 动态修改，如 binlog_cache_size Mysql> set global 等。

有些参数既支持 session 也支持 global，如 read_buffer_size。

session 和 global 所表示的含义不一样：session 表示只在当前的会话中生效，针对其他会话时不生效。global 表示全局生效，对所有登录的会话都生效，如要退出，需重新登录才生效。

2. 说明

（1）在 MySQL 8.0 中，若要 set global 参数需要权限 SYSTEM_VARIABLES_ADMIN or SUPER。

（2）在配置文件中有很多开关，此时设置为数字 1、true 或 on 的效果一样都表示开启，相反的 0、false、off 都表示关闭。建议开启和关闭使用 on 和 off。

（3）在 MySQL8.0 中，修改了全局的变量值，只能在 MySQL 实例的生命周期内生效，且不能对 MySQL 的参数文件修改。MySQL 实例重启则无效，若要生效，需要将参数文件写到配置文件等下次重启之后生效。

5.1.3 参数持久化

MySQL 的动态参数可以在运行时通过 SET GLOBAL 命令来更改，但是这种更改只会临时生效，到下次启动时数据库又会从配置文件中读取。MySQL8.0 之前的参数文件的动态修改不能写到配置文件，MySQL8.0 支持了动态修改参数并将其保存到一个新的参数文件中 mysqld - auto.cnf，并默认保存在 $datadir 目录下。在所有的参数之后启动，需要的权限为 SYSTEM_VARIABLES_ADMIN or SUPER。

mysql> set persist max_connections= 300;

可以使用数据字典查询。

mysql> SELECT * FROM performance_schema.persisted_variables;

对于已经持久化了变量,可通过 reset persist 命令清除掉,其只是清空 mysqld-auto.cnf 和 performance_schema.persisted_variables 中的内容。

MySQL 会将该命令的配置保存到数据目录下的 mysqld-auto.cnf 文件中,下次启动时会读取该文件,用其中的配置来覆盖缺省的配置文件。不建议手动修改该文件,因为其内容是 json 格式的,在启动过程中极有可能导致数据库因解析错误而失败,因为有些参数不支持持久化的。

5.1.4 参数文件常用配置

1. Windows 系统

```
[mysqld]
port=3306
basedir=E:\mysql-8.0.26-winx64
datadir=E:\mysql-8.0.26-winx64\data80263307
max_connections=200
max_connect_errors=10
character-set-server=utf8mb4
default-storage-engine=INNODB
default_authentication_plugin=mysql_native_password
```

2. Linux 系统

```
[mysqld]
mysqld=/usr/local/mysql80/mysql8019/bin/mysqld_safe
mysqladmin=/usr/local/mysql80/mysql8019/bin/mysqladmin
port=3509
basedir=/usr/local/mysql80/mysql8019
datadir=/usr/local/mysql80/mysql8019/data
socket=/usr/local/mysql80/mysql8019/data/mysqls80193509.sock
default_authentication_plugin=mysql_native_password
server_id=80193509
log-bin
default-time-zone = '+8:00'
log_timestamps = SYSTEM
skip-name-resolve
```

character_set_server= utf8mb4

5.1.5 大小写敏感参数

MySQL 中操作系统对大小写的敏感性决定了数据库和表的大小写敏感，这里将自己遇到的坑分享给大家。之前接了一个项目安排团队成员开发，在自己本机 Windows 上开发和测试过程中一直没有问题，临近项目交付的时候，部署到 Linux 服务器上后，发现有报错，日志信息大概是：

MySQLSyntaxErrorException：Table 'mes_db.student' doesn't exist

出现了问题，怎么部署服务器就不行了。错误提示很明显，student 表不存在！

于是打开 navicat，查看这个表在不在，一看还真在，数据库中显示的 student，不过 s 是小写。

查看代码，发现代码中发现把表名写成 Student，就一个 s 写成大写 S 了。

问题找到了，原来是不小心写 SQL 的时候没有写对表名，改一下表名就搞定了，功能也一切正常了。请大家想这个问题，为什么本地 Window 环境就一直没有出现这个报错提示呢？非要部署到 Linux 服务器才出现，这到底是什么原因？

经过排查发现，MySQL 在 Linux 下默认是区分大小写的，由 lower_case 参数控制。

当 lower_case_table_names 为 0 时表示区分大小写，为 1 时表示不区分大小写。

0——大小写敏感（Unix，Linux 默认）

1——大小写不敏感（Windows 默认）

2——大小写不敏感（macOS 默认）

其实，官网也提示说：如果在数据目录驻留不区分大小写的文件系统（例如 Windows 或 macOS）上运行 MySQL，则不应将 lower_case_table_names 设置为 0。后来在 window10 环境尝试设置 lower_case_table_names 为 0 的时候，MySQL 的服务怎么也启动不了，启动服务报错，因为 windows 系统对大小写不敏感。

总结经验，操作系统不同导致大小写敏感不一致，所以我们在开发时，应该按大小写敏感的原则去开发，这样可以使开发的程序兼容不同的操作系统。因此，建议在开发测试环境下把 lower_case_table_names 的值设为 0，便于在开发中就严格控制代码大小写问题，以提高代码的兼容和严谨。

5.2 日志文件

日志文件记录了影响 MySQL 数据库各种类型的活动，用来记录数据库的运行情况、日常操作和错误等信息，可以帮助我们诊断数据库出现的各种问题，常见的日志文件有错误日志、二进制日志、慢查询日志、全查询日志、redo 日志、undo 日志等。

5.2.1 错误日志

错误日志(Error Log)是 MySQL 中最常用的一种日志,主要记录 MySQL 服务器启动和停止过程中的信息、服务器在运行过程中发生的故障和异常情况等。MySQL DBA 在遇到问题时候,第一时间应该查看这个错误日志文件,该文件不但记录了出错信息,还记录了一些警告信息以及正确信息,这个 error 日志文件类似于 oracle 的 alert 文件。错误日志以文本文件的形式存储,直接使用普通文本工具就可以查看,默认情况下是以 err 结尾。

在 MySQL 中,通过 SHOW 命令可以查看错误日志文件所在的目录及文件名信息。

```
mysql> SHOW VARIABLES LIKE 'log_error'
Variable_name    Value
log_error        G:\mysql-8.0.23-winx64\data80323308\error.err
1 row in set, 1 warning (0.00 sec)
```

1. 错误日志配置

默认是启动的,一般以 err 做后缀名,需要在参数文件中配置,将 log_error 选项加入到 MySQL 配置文件的[mysqld]组中,形式如下:

```
[mysqld]
log_error = /usr/local/mysql/data/error.err
```

2. 错误日志清理

```
[root@jeames data]# mv log.err log-old.err
[root@jeames ~ ]# mysqladmin -uroot -p flush-logs
```

此时应通过 mysqladmin 命令来开启新的错误日志,以保证 MySQL 服务器上的硬盘空间。

5.2.2 慢查询日志

慢查询日志(slow_query_log)主要用来记录执行时间超过设置的某个时长的 SQL 语句,能够帮助数据库维护人员找出执行时间比较长、执行效率比较低的 SQL 语句,并对这些 SQL 语句进行针对性优化。

慢查询日志可以帮助 DBA 找出执行效率缓慢的 SQ 语句,为数据库优化工作提供帮助。

慢查询日志默认是不开启的,建议开启慢查询日志。

当需要进行采样分析时,手工开启。

1. 配置慢查询日志

mysql> SHOW VARIABLES LIKE 'log_err';

可以使用以上命令查询慢查询日志配置,参数文件中配置开启慢查询日志,重启 MySQL 服务即可。

[mysqld]
slow_query_log = 1
slow_query_log_file = /data/mysql/log/query_log/slow_statement.log
long_query_time = 10
log_output = FILE

各配置项说明如下:

slow_query_log:指定是否开启慢查询日志,1 表示开启慢查询日志,0 表示关闭慢查询日志。

slow_query_log_file:慢查询日志的文件位置。

long_query_time:指定 SQL 语句执行时间超过多少秒时记录慢查询日志。

log_output:支持将日志记录写入文件,也支持将日志记录写入数据库表,FILE 表示记录到文件,TABLE 表示记录到表,当记录到数据表中时,慢查询时间只能精确到秒;如果是记录到日志文件中,则慢查询时间能够精确到微秒,建议在实际工作中,将慢查询日志记录到文件中。

除了在文件中配置开启慢查询日志外,也可以在 MySQL 命令行中修改参数开启慢查询日志。

mysql> SET GLOBAL slow_query_log = 1
mysql> SET GLOBAL slow_query_log_file = '/data/mysql/log/query_log/slow_statement.log'
mysql> SET GLOBAL long_query_time = 10
mysql> SET GLOBAL log_output = 'FILE'

可以使用如下命令测试慢查询日志:

mysql> select sleep(10)
[root@jeames ~]# tail -n 5 /var/lib/mysql/jeames-slow.log
Time: 2022-09-23T23:57:06.096327Z
User@Host: skip-grants user[root] @ localhost [] Id: 7
Query_time: 10.000994 Lock_time: 0.000000 Rows_sent: 1 Rows_examined: 1
SET timestamp= 1663977416

2. 慢查询日志管理

慢查询日志和查询日志一样以纯文本文件的形式存储在服务器磁盘中,可以直接删除。

(1) 删除慢查询日志

```
mysql> SET GLOBAL slow_query_log = 'OFF'
[root@jeames ~]# ll /var/lib/mysql/jeames-slow.log
-rw-r----- 1 mysql mysql 394 Sep 24 07:57 /var/lib/mysql/jeames-slow.log
[root@jeames ~]# rm -rf /var/lib/mysql/jeames-slow.log
mysql> SET GLOBAL slow_query_log = 'ON'
mysql> select @@slow_query_log
+------------------+
| @@slow_query_log |
+------------------+
|                1 |
+------------------+
mysql> flush slow logs;
```

此时我们就能看到新创建的慢查询日志

```
[root@jeames ~]# du -sh /var/lib/mysql/jeames-slow.log
4.0K    /var/lib/mysql/jeames-slow.log
```

当关闭慢查询日志后，删除慢查询日志文件，再执行刷新日志的操作，MySQL将不再重新创建慢查询日志文件。

(2) 慢查询日志写到表里

`log_output` 默认是 `FILE`，表示慢查询日志输入至日志文件，可以通过 set 修改输出为 `TABLE`。

```
mysql> show variables like '%log_output%'
+---------------+-------+
| Variable_name | Value |
+---------------+-------+
| log_output    | FILE  |
+---------------+-------+
1 row in set (0.01 sec)
mysql> SET GLOBAL slow_query_log = 'OFF'
```

--修改参数，慢查询日志输入至表中

```
mysql> set global log_output = 'TABLE'
mysql> use mysql
```

--清空慢查询日志表
mysql> RENAME TABLE slow_log TO slow_log_temp
mysql> DELETE FROM slow_log_temp WHERE start_time < DATE(NOW())
mysql> RENAME TABLE slow_log_temp TO slow_log

--打开慢查询日志
mysql> SET GLOBAL slow_query_log = 'ON'
--确认修改是否生效
mysql> select * from performance_schema. global_variables where variable_name in
    -> ('slow_query_log','log_output','slow_query_log_file','long_query_time')

```
+--------------------+-------------------------------+
| VARIABLE_NAME      | VARIABLE_VALUE                |
+--------------------+-------------------------------+
| log_output         | TABLE                         |
| long_query_time    | 10.000000                     |
| slow_query_log     | ON                            |
| slow_query_log_file| /var/lib/mysql/jeames-slow.log|
+--------------------+-------------------------------+
```

3. 分析工具 mysqldumpslow

我们通过查看慢查询日志可以发现，当很乱或数据量大的时候，可能一天会产生几个 G 的日志，根本没有办法进行清晰明了的分析，所以这里我们采用 MySQL 自带的慢查询日志分析工具分析日志 mysqldumpslow 工具。

(1)查看 mysqldumpslow 的帮助信息

[root@jeames ~]# mysqldumpslow --help

-s:表示按何种方式排序：

c:访问次数

l:锁定时间

r:返回记录

t:查询时间

al:平均锁定时间

ar:平均返回记录数

at:平均查询时间

-t:返回前面多少条的数据

-g:后边搭配一个正则匹配模式,大小写不敏感的

(2)常用命令

得到返回记录集最多的 10 个 SQL：

mysqldumpslow –s r –t 10 /var/lib/mysql/show.log

得到访问次数最多的 10 个 SQL：

mysqldumpslow –s c –t 10 /var/lig/mysql/show.log

得到按照时间排序的前 10 条里面含有左连接的查询语句：

mysqldumpslow –s t –t 10 –g"left join" /var/lig/mysql/show.log

另外，建议在使用这些命令时，结构 | 和 more 同时使用，否则有可能出现爆屏情况：

mysqldumpslow –s r –t 10 /var/lig/mysql/show.log | more

5.2.3　全查询日志

全查询日志记录了所有对数据库请求的信息，正确的 SQL 才会被记录下来（错误写法的 SQL 语句不会记录），包括 show、查询 select 语句、权限不足的语句（ERROR 1044（42000）：Access denied for user）。默认位置在变量 datadir 下，默认文件名为主机名.log。

MySQL 的通用查询日志在默认情况下是不开启的，当需要进行采样分析时，可手工开启。

1. 开启全查询日志

可以在 MySQL 命令行中修改参数，开启全查询日志

mysql> SET GLOBAL general_log = 1

mysql> SET GLOBAL general_log_file = '/data/mysql/log/general_statement.log'

mysql> SET GLOBAL log_output = 'FILE';

注意：查询日志记录查询语句与启动时间，建议不是在调试环境下不要开启查询日志，因为会不断占据你的磁盘空间，并会产生大量的 IO。

2. 全查询日志管理

(1)日志清理

--关闭全查询日志

mysql> SET GLOBAL general_log = 'OFF'

--备份文件

[root@jeames data]# mv general_statement.log general_statement_old.log

--重新生成全查询日志

mysql> flush general logs

也可以在服务器命令行中执行如下命令刷新日志。

[root@jeames ~]# mysqladmin –uroot –p flush–logs

（2）全查询日志表清理
```
mysql> SET GLOBAL general_log = 'OFF'
mysql> RENAME TABLE general_log TO general_log_temp
mysql> DELETE FROM general_log_temp WHERE event_time < DATE(NOW())
mysql> RENAME TABLE general_log_temp TO general_log
mysql> SET GLOBAL general_log = 'ON'
```

注意：全查询日志支持日志写入文件或者表中，通过修改 log_output，可以通过 set 修改输出为 TABLE。

5.2.4 二进制日志

二进制日志中以"事件"的形式记录了数据库中数据的变化情况，对于 MySQL 数据库的灾难恢复起着重要的作用。其记录了对数据库进行变更的操作，但不包括 select 操作以及 show 操作，因为这类操作对数据库本身没有修改。如果想记录 select 和 show，那就需要开启全查询日志，binlog 还包括了执行数据库更改操作时间和执行时间等信息。

在 MySQL8.0 中，二进制日志默认开启，需在 MySQL 参数文件中配置慢查询日志。

--参数查询
```
mysql> show variables like '%log_bin%'
```
--参数文件如下配置
```
[mysqld]
log_bin = ON
log_bin_basename = /var/lib/mysql/binlog
log_bin_index = /var/lib/mysql/binlog.index
server_id= 80233306
binlog_format= mixed
binlog_cache_size= 32m
max_binlog_cache_size= 64m
max_binlog_size= 512m
# expire_logs_days = 10   mysql 8 开始 expire_logs_days 废弃
binlog_expire_logs_seconds = 2592000 # 秒为单位
```

各项配置说明如下：

log_bin：是否启用二进制日志

log_bin_basename：指定了 binlog 的基础命名和存储路径，文件命名方式为 mysql-bin.0000XX log_bin_index：这个文件不记录二进制内容，存储了启用的 binglog 完整文件名

server_id：在 MySQL8.0 版本中，若想开启二进制日志，则必须加上 server_id 参数

binlog_format：二进制文件的格式，取值可以是 STATEMENT、ROW 和 MIXED

STATEMENT 格式表示二进制日志文件记录的是日志的逻辑 SQL 语句。

优点：不需要记录每一行的数据变化，减少 bin-log 日志量，节约磁盘 IO，提高性能。

缺点：容易出现主从复制不一致的问题。

在 ROW 格式下，二进制日志记录的不再是简单的 SQL 语句了，而是记录表的行更改情况，此时可以将 InnoDB 的事务隔离基本设为 READ COMMITTED，以获得更好的并发性。通过 binlog 获取历史的 SQL 执行记录，前提是必须开启 binlog_rows_query_log_events 参数。

优点：能清楚记录每一行数据修改的细节。

缺点：数据量太大。

在 MIXED 格式下，MySQL 默认采用的 STATEMENT 格式进行二进制日志文件的记录，但是在一些情况下会使用 ROW 格式，可能的情况包括：

表的存储引擎为 InnoDB，这时对于表的 DML 操作都会以 ROW 格式记录

使用了 UUID()、USER()、CURRENT_USER()、FOUND_ROWS()、ROW_COUNT()等不确定函数

使用了 INSERT DELAY 语句

使用了用户定义函数

使用了临时表

说明：binlog_cache_size：二进制日志的缓存大小，如果经常有大事务操作，可以将这个值调大一点点。

max_binlog_cache_size：对于二进制日志的最大缓存大小，根据经验最大值推荐 4GB

max_binlog_size：单个二进制日志文件的最大大小，当文件大小超过此选项配置的值时，会发生日志滚动，重新生成一个新的二进制文件，最大值是 1 GB。

binlog_expire_logs_seconds：二进制日志的过期时间。如果配置了此选项，则 MySQL 会自动清理过期的二进制日志。此选项的默认值为 0，表示 MySQL 不会清理过期日志。配置完成后，重启 MySQL 才能使配置生效，此时，会在 log_bin 指定的目录下生成 MySQL 的二进制文件。

其他：

binlog_do_db 和 binlog_ignore_db

binlog_do_db = db1 #此参数表示只记录指定数据库的二进制日志，默认全部记录

binlog_do_db = db2

```
binlog_ignore_db = db3 #此参数表示不记录指定的数据库的二进制日志
binlog_ignore_db = db4
```

binlog 是 MySQL Server 层记录的二进制日志文件,常用命令如下：

--查询二进制日志配置信息

```
mysql> show variables like '%log_bin%'
+---------------------------------+------------------------+
| Variable_name                   | Value                  |
+---------------------------------+------------------------+
| log_bin                         | ON                     |
| log_bin_basename                | /var/lib/mysql/binlog  |
| log_bin_index                   | /var/lib/mysql/binlog.index |
| log_bin_trust_function_creators | OFF                    |
| log_bin_use_v1_row_events       | OFF                    |
| sql_log_bin                     | ON                     |
+---------------------------------+------------------------+
```

--滚动日志,切日志

```
mysql> flush logs
```

--查询二进制日志

```
mysql> show binary logs
mysql> show master logs
+---------------+-----------+-----------+
| Log_name      | File_size | Encrypted |
+---------------+-----------+-----------+
| binlog.000012 |       156 | No        |
| binlog.000013 |       179 | No        |
| binlog.000014 |       179 | No        |
| binlog.000015 |       179 | No        |
| binlog.000016 |       156 | No        |
| binlog.000017 |       156 | No        |
| binlog.000018 |       200 | No        |
| binlog.000019 |       200 | No        |
| binlog.000020 |       358 | No        |
| binlog.000021 |       746 | No        |
| binlog.000022 |       200 | No        |
| binlog.000023 |       156 | No        |
+---------------+-----------+-----------+
```

12 rows in set (0.06 sec)

--查询二进制当前日志

```
mysql> show master status
+-------------+----------+--------------+------------------+-------------------+
| File        | Position | Binlog_Do_DB | Binlog_Ignore_DB | Executed_Gtid_Set |
+-------------+----------+--------------+------------------+-------------------+
| binlog.000023|    156  |              |                  |                   |
+-------------+----------+--------------+------------------+-------------------+
1 row in set (0.00 sec)
```

--查看二进制日志记录的事件(event)

```
mysql> show binlog events in 'binlog.000022'
+--------------+-----+--------------+-----------+------------+------------------------------+
| Log_name     | Pos | Event_type   | Server_id | End_log_pos| Info                         |
+--------------+-----+--------------+-----------+------------+------------------------------+
| binlog.000022|   4 | Format_desc  |         1 |        125 | Server ver. 8.0.27, Binlog ver. 4 |
| binlog.000022| 125 | Previous_gtids|        1 |        156 |                              |
| binlog.000022| 156 | Rotate       |         1 |        200 | binlog.000023;pos=4          |
+--------------+-----+--------------+-----------+------------+------------------------------+
3 rows in set (0.00 sec)
```

二进制日志文件不能以纯文本文件的形式来查看,可以使用MySQL的mysql-binlog命令进行查看

```
[root@jeames ~]# mysqlbinlog --no-defaults binlog.000023
```

日常的运维中,binlog日志增长大,会造成服务器磁盘爆满,清理二进制日志方法如下:

自动清理:

启用binlog_expire_logs_seconds设置binlog自动清除日志时间

```
mysql> show variables like '%binlog_expire_logs_seconds%'
+----------------------------+---------+
| Variable_name              | Value   |
+----------------------------+---------+
| binlog_expire_logs_seconds | 2592000 |
+----------------------------+---------+

mysql> set global binlog_expire_logs_seconds=86400
```

注意:保存时间以秒为单位,86400表示1天,MySQL会自动清理过期的二进制

日志。

手动清理：

默认日志文件达到 1G 都会重新生成一个新的二进制日志文件。

mysql> select @@max_binlog_size

-- binlog.000025 之前的日志都会被删除

mysql> PURGE BINARY LOGS TO 'binlog.000025'

-- 时间'2022-09-28 23:59:59'之前的日志都会被删除

mysql> PURGE BINARY LOGS BEFORE '2022-09-28 23:59:59'

-- 重置历史二进制日志，从 000001 开始重新

mysql> RESET MASTER

5.2.5 Redo 日志

1. Redo 日志概念及特点

Redo 日志也叫事务日志，是记录 InnoDB 等支持事务的存储引擎执行事务时产生的日志。事务型存储引擎用于保证日志的原子性、一致性、隔离性和持久性。注意数据库变更数据不会立即写到数据文件中，而是写到事务日志中。

所有的数据库都是日志先行，先写日志，再写数据文件，所以才会有 redo log 的规则。在默认情况下，会出现 2 个文件名称分别为 ib_logfile0 和 ib_logfile1，在 MySQL 数据库目录下可以看到这 2 个文件，这对 innodb 存储引擎非常重要，因为它们记录了对于 innodb 存储引擎的事务日志，是物理层面的。

其实它的作用是在 MySQL 宕机之后，用来恢复数据的一种 log 文件，其里面包含了表空间、数据页号、磁盘文件偏移量、更新值、日志类型这些信息，其次 redo log 是按顺序写的，所以每次都是追加到磁盘文件末尾，速度非常快。

每个 redo log 都是按顺序写入到 redo log block 中，一个 redo log block 最多存放 496 字节的 redo log，当一个 redo log block 被写满之后，再将这个 redo log block 写入 redo log 日志文件中，在这里，其实还有一个概念，就是 redo log buffer，顾名思义，是一个缓存，类似于 Buffer Pool，这里面缓存的是多个空的 redo log block。只有满足了一定的条件，redo log buffer 才会真正将我们的 redo log 刷入磁盘中。

2. redo log block 真正刷入磁盘的时机

一个 redo log buffer 默认是 16 M，redo log buffer 的日志占了总容量的一半，一半也就是 8 MB，就刷盘。

当一个事务提交的时候，redo log 所在的 redo log block 就会刷盘。

后台线程每隔一秒将 redo log buffer 里面的 redo log block 刷入磁盘。

MySQL 关闭的时候，redo log block 刷盘。

(1)一旦有事务提交，会立即写入此文件，一般查看参数设置如下：

```
mysql> show variables like 'innodb%log%'
+---------------------------------------+------------+
| Variable_name                         | Value      |
+---------------------------------------+------------+
| innodb_api_enable_binlog              | OFF        |
| innodb_flush_log_at_timeout           | 1          |
| innodb_flush_log_at_trx_commit        | 1          |
| innodb_log_buffer_size                | 16777216   |
| innodb_log_checksums                  | ON         |
| innodb_log_compressed_pages           | ON         |
| innodb_log_file_size                  | 50331648   |
| innodb_log_files_in_group             | 2          |
| innodb_log_group_home_dir             | ./         |
| innodb_log_spin_cpu_abs_lwm           | 80         |
| innodb_log_spin_cpu_pct_hwm           | 50         |
| innodb_log_wait_for_flush_spin_hwm    | 400        |
| innodb_log_write_ahead_size           | 8192       |
| innodb_log_writer_threads             | ON         |
| innodb_max_undo_log_size              | 1073741824 |
| innodb_online_alter_log_max_size      | 134217728  |
| innodb_print_ddl_logs                 | OFF        |
| innodb_redo_log_archive_dirs          |            |
| innodb_redo_log_encrypt               | OFF        |
| innodb_undo_log_encrypt               | OFF        |
| innodb_undo_log_truncate              | ON         |
+---------------------------------------+------------+
21 rows in set (0.00 sec)
```

(2)innodb_flush_log_at_trx_commit 参数

默认值 1 的意思是每一次事务提交或事务外的指令都需要把日志写入（flush）硬盘，这是很费时的。

设成 2 对于生产上应用很多，它的意思是不写入硬盘而是写入系统缓存，日志仍然会每秒 flush 到硬盘，所以一般不会丢失超过 1~2 t 的更新。

(3)Redo 与 Binlog 区别

所有的数据库都是日志先行，先写日志，再写数据文件，所以才会有 redo log 的规则。

为了保证事务的持久性，mysql 的 InnoDB 采用了 WAL 技术，WAL 的全称是 Write-Ahead Logging。

① Redo Log：Redo Log 是 InnoDB 存储引擎提供的一种物理日志结构，用来描述对底层数据页操作的具体内容，记录物理页的修改，主要用于实现崩溃恢复（crash-recover），并提升磁盘操作效率。

② Binlog：Binlog 是 MySQL Server 本身提供的一种逻辑日志，其与具体存储引擎无关，描述的是数据库所执行的 SQL 语句或数据变更情况，主要用于数据复制和增量恢复。

5.2.6　undo 日志

Undo 日志是为了实现事务的原子性，在 MySQL 数据库 InnoDB 存储引擎中，还用 Undo 日志来实现多版本并发控制（简称 MVCC）。

undo 日志是与单个读写事务关联的撤销日志记录的集合。undo 日志记录包含有关如何撤消事务对聚集索引记录的最新更改的信息。如果另一个事务需要将原始数据作为一致读取操作的一部分，则从 undo 日志记录中检索未修改的数据。undo 日志存在于 undo 日志段中，该段包含在回滚段中。回滚段驻留在 undo 表空间和全局临时表空间中。

Undo 日志的原理很简单，为了满足事务的原子性，在操作任何数据之前，首先将数据备份到一个地方（这个存储数据备份的地方称为 Undo Log）。然后进行数据的修改。如果出现了错误或者用户执行了 ROLLBACK 语句，系统可以利用 Undo Log 中的备份将数据恢复到事务开始之前的状态。

除了可以保证事务的原子性，Undo Log 也可以用来辅助完成事务的持久化。

每个 undo 表空间和全局临时表空间分别支持最多 128 个回滚 innodb_rollback_segments 变量定义回滚段的数量。一个事务最多分配 4 个 undo 日志，分别用于以下操作类型：

(1) 用户定义表上的 INSERT 操作。
(2) 用户定义表上的 UPDATE 和 DELETE 操作。
(3) 对用户定义的临时表执行 INSERT 操作。
(4) 对用户定义的临时表执行 UPDATE 和 DELETE 操作。

5.2.7　中继日志

如图 5-1 所示，从数据库 Slave 服务的 I/O 线程，即从主数据库 Master 服务的二进制日志中读取数据库的更改记录，并写入到中继日志中，然后在 Slave 数据库执行修改操作，这就是中继日志 Relay Log。

基本原理流程中 3 个线程以及之间的关联如下：

(1) Binary log：主数据库的二进制日志

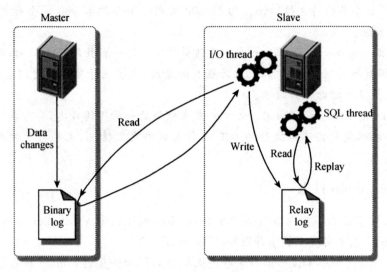

图 5-1　MySQL 主从复制工作原理

(2) Relay log：从服务器的中继日志

第一步：master 在每个事务更新数据完成之前，将该操作记录串行地写入 binlog 文件中。

第二步：salve 开启一个 I/O Thread，该线程在 master 打开一个普通连接，主要工作是 binlog dump process。如果读取的进度已经跟上了 master，就进入睡眠状态，并等待 master 产生新的事件。I/O 线程最终的目的是将这些事件写入到中继日志中。

第三步：SQL Thread 会读取中继日志，并按顺序执行该日志中的 SQL 事件，从而与主数据库中的数据保持一致。

5.3　MySQL 表结构文件

　　InnoDB 中用于存储数据的文件共有两个部分，一是系统表空间文件，包括 ibdata1、ibdata2 等文件，其中存储了 InnoDB 系统信息和元数据，是所有表公用的。另一个是 .idb 文件，是每张表独有的。.ibd 文件就是每一个表独有的表空间，文件存储了当前表的数据，索引数据和插入缓冲等信息。

　　innodb 包括 ibd 和 frm，当启用了 innodb_file_per_table，表被存储在他们自己的表空间里，由 innodb_file_per_table 参数控制。

```
mysql> show variables like 'innodb_file_per_table'
+-----------------------+-------+
| Variable_name         | Value |
+-----------------------+-------+
```

| innodb_file_per_table | ON |
+-------------------+-----+

当 MySQL8.0 以后没有 .frm 了，元数据都存在系统表空间里 ibdata1、ibdata2。

5.4 其他文件

5.4.1 pid 文件

当 mysql 实例启动的时候，会将自己的进程 id 写入一个文件中，该文件即为 pid 文件，由参数 pid_file 控制，默认路径位于数据目录下，大家可以通过以下 2 种方式查看：

```
mysql> show variables like 'pid_file';
+---------------+----------------------------+
| Variable_name | Value                      |
+---------------+----------------------------+
| pid_file      | /var/run/mysqld/mysqld.pid |
+---------------+----------------------------+
1 row in set (0.00 sec)

mysql> select @@pid_file;
+----------------------------+
| @@pid_file                 |
+----------------------------+
| /var/run/mysqld/mysqld.pid |
+----------------------------+
1 row in set (0.00 sec)
```

5.4.2 socket 文件

在 Linux 系统中本地连接 MySQL 可以采用 Linux 域套接字 socket 方式，需要一个套接字 socket 发文件，由参数 socket 控制。如果 socket 文件丢失，那么会导致不能从本地登录 MySQL，可以通过重启的方式来重新生成，通过如下 2 种方式查看位置：

1. 操作系统查找

ps – ef | grep mysql

2. 命令行查找

mysql> show variables like 'socket'

mysql> select @@socket

如果服务器有多个 MySQL,那么如何通过 socket 登录 MySQL 呢？方法如图 5-2 所示：

[root@jeames ~]# mysql -uroot -p -S /tmp/mysql.sock

```
[root@jeames ~]# mysql -uroot -p -S /tmp/mysql.sock
Enter password:
Welcome to the MySQL monitor.  Commands end with ; or \g.
Your MySQL connection id is 10
Server version: 8.0.19 MySQL Community Server - GPL

Copyright (c) 2000, 2020, Oracle and/or its affiliates. All rights reserved.

Oracle is a registered trademark of Oracle Corporation and/or its
affiliates. Other names may be trademarks of their respective
owners.

Type 'help;' or '\h' for help. Type '\c' to clear the current input statement.
```

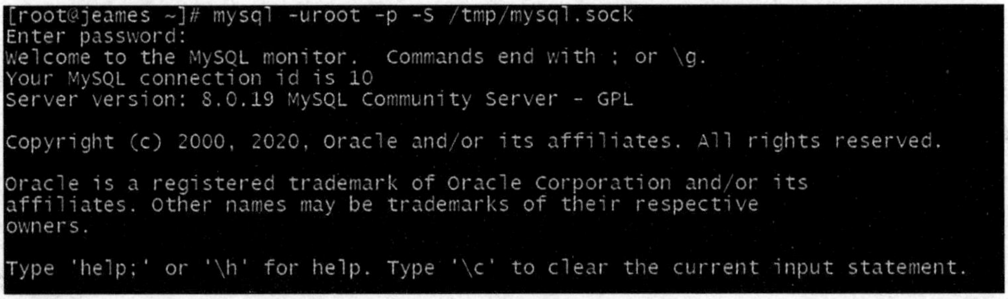

图 5-2 socket 登陆 MySQL

mysql.sock 是 MySQL 的主机和客户机在同一 host 上时,使用 unix domain socket 作为通信协议的载体,它比 tcp 快。对 mysql.sock 来说,其作用是程序与 mysqlserver 处于同一台机器,发起本地连接时可用。

第 6 章 存储引擎结构

存储引擎是对底层物理数据执行实际操作的组件，为 Server 服务器层提供各种操作数据的 API，数据是被存放在内存或者是磁盘中的。MySQL 支持插件式的存储引擎，包括 InnoDB、MyISAM、Memory 等。一般情况下，MySQL 默认使用的存储引擎是 InnoDB。如图 6-1 所示，InnoDB 存储引擎整体分为内存架构（Memory Structures）和磁盘架构（Disk Structures）。

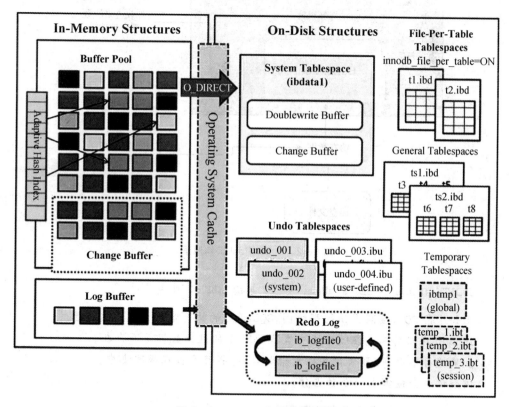

图 6-1　InnoDB 存储引擎架构

6.1　内存结构

内存结构主要包括 Buffer Pool、Change Buffer、Adaptive Hash Index 和 Log Buffer 4 大组件。

6.1.1 Buffer Pool

1. Buffer Pool 原理

Buffer Pool(缓冲池)是 InnoDB 存储引擎中非常重要的内存结构,类似 Redis 一样的作用,原理如图 6-2 所示,起缓存的作用。MySQL 的数据最终是存储在磁盘中的,如果没有 Buffer Pool,那么每次的数据库请求都会磁盘中查找,这样必然会存在 IO 操作。但是有了 Buffer Pool,在第一次查询时就会将查询的结果保存到 Buffer Pool 中,这样后面再有请求时就会先从缓冲池中查询,如果没有再去磁盘中查找,然后再放到 Buffer Pool 。

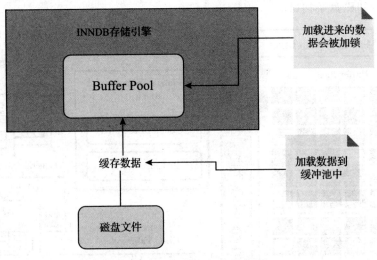

图 6-2 Buffer Pool 工作原理

```
UPDATE students SET stuName = 'IT 邦德' WHERE id = 1
```

比如这条 SQL,按照上面的那幅图,SQL 语句的执行步骤大致是这样子的。

innodb 存储引擎先在缓冲池中查找 id=1 的这条数据是否存在。

如果缓存不存在,那么就去磁盘中加载,并将其存放在缓冲池中。

该条记录会被加上一个独占锁。

2. buffer pool 和查询缓存的区别

(1)查询缓存:查询缓存位于 Server 层,MySQL Server 首选会从查询缓存中查看是否曾经执行过这个 SQL,如果曾经执行过,之前执行的查询结果会以 Key-Value 的形式保存在查询缓存中。key 是 SQL 语句,value 是查询结果。我们将这个过程称为查询缓存!

(2)Buffer Pool 位于存储引擎层,Buffer Pool 就是 MySQL 存储引擎为了加速数据的读取速度而设计的缓冲机制,为了提高缓存管理的效率,使用页面链表的方式+LRU(最近最少使用)算法进行管理。该区域大小由参数 innodb_buffer_pool_size 来定义,默认 128 MB。

Buffer Pool 以 Page 页为单位,默认大小 16K,BP 的底层采用链表数据结构管理 Page。在 InnoDB 访问表记录和索引时会在 Page 页中缓存,以后使用可以减少磁盘 IO 操作,提升效率,可以将 80% 的内存分配给 InnoDB 缓冲池。

其实 MySQL 会有一个后台线程,它会在某个时机将我们 Buffer Pool 中的脏数据刷到 MySQL 数据库中,这样就将内存和数据库的数据保持统一了。

刷新脏页的时机如下:

redo log 写满时,停止所有更新操作,将 checkpoint 向前推进,推进那部分日志的脏页更新到磁盘。

需要将一部分数据页淘汰,如果是干净页,直接淘汰就行了,脏页需要全部同步到磁盘。

MySQL 自认为空闲时。

MySQL 正常关闭之前。

6.1.2 Change Buffer

Insert buffer part of buffer pool 是一种特殊的数据结构(早期只支持 INSERT 操作的缓冲,所以也叫作 Insert Buffer),当受影响的页面不在缓冲池中时,将会缓存对辅助索引页的更改。这些更改可能是由 INSERT、UPDATE、DELETE(DML)语句执行时所导致的,当其他读取操作从磁盘中加载数据页时,如果这些数据页包含 Change Buffer 中缓存的更改操作页,那么将进行合并操作。

1. Change Buffer 原理:

(1)在需要更新一个数据页时,如果数据页在 buffer pool 内存中就直接更新。如果数据页不在 buffer pool 内存中,在不影响数据一致性的前提下,innodb 会将这些更新操作缓存在 change buffer 中,这样就不需要从磁盘中读入这个数据页了。在下次查询需要访问这个数据页的时候,将数据页读入 buffer pool 内存中,然后执行 change buffer 中与这个页有关的操作。通过这种方式就能保证这个数据逻辑的正确性。

(2)将 change buffer 中的操作合并到原数据页,得到最新结果的过程称为 merge,以下情况会触发 merge。

① 访问这个数据页。

② 后台 master 线程会定期 merge。

③ 数据库缓冲池不够用时。

④ 数据库正常关闭时。

⑤ redo log 写满时。

2. Change Buffer 好处：

将数据从磁盘读入内存涉及随机 IO 的访问是数据库里面成本较高的操作。change buffer 因为减少了随机磁盘访问，所以对更新性能的提升是很明显的。写多读少，如果所有的更新后，都马上伴随着对这个记录的查询，那么应该关闭 change buffer。而在其他情况下，change buffer 都能提升更新性能。

3. Change Buffer 的相关参数

（1）innodb_change_buffering；对哪个操作数据需要更新到 change buffer。

innodb_change_buffering = all   //默认是 all
all   //所有操作 包括 inserts deletes purges
none   //不需要
inserts //插入
deletes //当数据被标记为删除时
purges   //当数据被物理删除时

（2）innodb_change_buffer_max_size；占用 buffer pool 多大比例，默认 25，最大 50。

Change Buffer 关系图如图 6-3 所示。

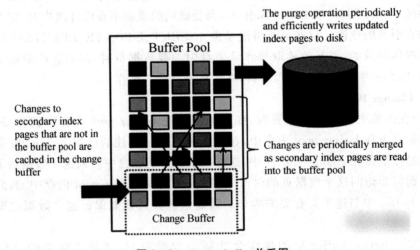

图 6-3 change buffer 关系图

6.1.3 Adaptive Hash Index

自适应哈希索引（AHI），可使 InnoDB 能够在具有适当工作负载组合和足够内存的系统上执行更新内存中数据库的操作，而不会牺牲事务特性或可靠性。

自适应哈希索引由 innodb_adaptive_hash_index 变量启用，或在服务器启动时由--skip innodb 自适应哈希指数关闭。

对于某些工作负载,哈希索引查找的速度大大超过了监视索引查找和维护哈希索引结构的额外工作。在繁重的工作负载(如多个并发联接)下,对自适应哈希索引的访问有时会成为争议源,即使使用 LIKE 运算符和％通配符的查询也不会有什么作用。对于无法从自适应哈希索引中受益的工作负载,关闭它可以减少不必要的性能消耗。由于很难预先预测自适应哈希索引是否适合特定的系统和工作负载,因此,推荐大家考虑在启用和禁用它的情况下运行基准测试。

6.1.4 Log Buffer

日志缓冲区(Log Buffer)是存储要写入磁盘上日志文件数据的内存区域。日志缓冲区大小由 innodb_Log_buffer_size 变量定义,默认大小为 16 MB。日志缓冲区的内容定期刷新到磁盘。大型日志缓冲区使大型事务能够运行,而无须在事务提交之前将重做日志数据写入磁盘。因此,如果您有更新、插入或删除许多行的事务,那么增加日志缓冲区的大小可以节省磁盘 I/O。

innodb_flush_log_at_trx_commit 变量控制如何将日志缓冲区的内容写入,并刷新到磁盘。innodb_flush_log_at_timeout 变量控制日志刷新频率。

例:update students set stuName = 'IT 邦德' where id = 1

到了这里,SQL 语句也更新好了,那么需要将更新的值提交了,也就是需要提交本次的事务,只要事务成功提交了,才会将最后的变更保存到数据库。在提交事务前将 redo Log Buffer 中的数据写入到磁盘文件 redo log 中,进行持久化,而刷入磁盘的时机则是根据 innodb_flush_log_at_trx_commit 参数来决定的:

0:默认值,每秒将 redo log buffer 中的数据将以写入到日志文件中,同时 flush 到磁盘。

1:在机器 crash 并重启后,会丢失一秒的事务日志数据 1:每次事务提交时,将 redo log buffer 中的数据写入日志文件,并同时 flush 到磁盘。在机器 crash 并重启后,不会丢失事务日志。

2:每次事务提交时,将 redo log buffer 中的数据写入日志文件,并每秒 flush 一次到磁盘。在机器 crash 并重启后,有可能丢失数据,因此,如果在应用场景中经常有大事务,则可以考虑增大重做日志缓冲区以减少磁盘 I/O 操作 。

1. bin log 和 redo log 区别

bin log 和 redo log 有些相似,两者的主要区别有:

(1)redo log 是 InnoDB 存储引擎特有的日志文件,而 bin log 属于 MySQL 级别的日志。

(2)redo log 适用于崩溃恢复,bin log 适用于主从复制和数据恢复。

(3)redo log 记录的东西是偏向于物理性质的,如"对什么数据,做了什么修改"。bin log 是偏向于逻辑性质的,类似于:"对 students 表中的 id 为 1 的记录做了更新操作"

bin log 文件是如何刷入磁盘的

mysql> show variables like '%sync%'

bin log 的刷盘策略可以通过 sync_bin log 来修改，默认为 0，表示先写入 os cache，也就是说在提交事务的时候，数据不会直接到磁盘中，这样如果宕机 bin log 数据仍然会丢失。所以建议将 sync_binlog 设置为 1，表示直接将数据写入到磁盘文件中。

2. bin log 在什么时候记录数据

其实 MySQL 在提交事务时，不仅会将 redo log buffer 中的数据写入到 redo log 文件中，也会将本次修改的数据记录到 bin log 文件中，同时会将本次修改的 bin log 文件名和修改的内容在 bin log 中的位置记录到 redo log 中，最后还会在 redo log 最后写入 commit 标记，这样就表示本次事务被成功地提交了。

6.2 磁盘结构

1. System Tablespace

InnoDB 系统表空间包含 InnoDB 数据字典（InnoDB 相关对象的元数据）。它之所以被称为系统表空间，是因为它可以被多个用户表共享。系统表空间可以由一个或多个数据文件构成。在默认情况下，只会创建一个名为 ibdata1 的共享表空间文件。但可以使用 innodb_data_file_path 启动选项控制共享表空间的数量和大小。

2. 双写

插入缓冲带来的是针对普通索引插入性能上的提升，而 double write 就是保证写入的安全性防止在 Mysql 实例发生宕机时，InnoDB 发生数据页部分页写（partialpage write）的问题。数据库实例崩溃，我们可以通过 redo log 进行恢复，不会有任何问题，但 redo log 文件记录的是页的物理操作，如果页都损坏了，是无法进行任何恢复操作的。所以我们需要页的一个副本，如果实例宕机了，可以先通过副本把原来的页还原出来，再通过 redo log 进行恢复，重做。这就是 double write 的作用。

双写系统空间中的存储区域（从 8.0.20 开始独立出来了），InnoDB 缓冲池中刷出的脏页在被写入数据文件之前，都先会写入 double write buffer，然后写缓冲区分 2 次，每次将 1MB 大小的数据写入磁盘共享表空间（double write），最后再从 double write buffer 写入数据文件。

3. Undo Logs

Undo Logs 用于存放事务修改之前的旧数据（undo log 记录了有关如何撤销事务

对聚集索引记录的最新更改的信息),基于 undo 实现了 MVCC 和一致性非锁定读。

4. File – Per – Table

设置参数 innodb_file_per_table=1 启用独立表空间时,每个表都会对应生成一个.ibd 文件,用于存放自己的索引和数据等;否则,在创建表时数据和索引将被存放在 ibdata 系统表空间文件中(系统表空间)。

5. General Tablespaces

使用 CREATE TABLESPACE 语法创建的 InnoDB 共享表空间(tablespace_name.ibd)。可以在 MySQLdatadir 之外创建,能够保存多个表,并支持所有行格式的表。

6. Undo Tablespace

undo 表空间包含一个或多个 undo log 文件,文件个数由配置参数 innodb_undo_tablespaces 控制。

7. Temporary Tablespace

临时表空间用于存放非压缩的 InnoDB 临时表和相关对象,其配置参数 innodb_temp_data_file_path 为临时表空间的数据文件定义了相对路径和初始大小等。innodb_temp_data_file_path 参数,则会在数据目录中创建一个名为 ibtmp1 的自动扩展的 12 MB 初始大小的文件。

8. Redo Logs

重做日志是在崩溃恢复期间使用的基于磁盘的数据结构文件,用于恢复不完整提交事务写入的数据。

6.3 MySQL 逻辑存储

InnoDB 逻辑存储单元主要分为表空间、段、区、页、行,如图 6-4 所示。

6.3.1 表空间

表空间用于存储表结构和数据,InnoDB 存储引擎表中所有数据都是存储在表空间中的,表空间分为系统表空间、独立表空间、通用表空间、临时表空间、Undo 表空间等。

1. 系统表空间(System Tablespace)

系统表空间包含 InnoDB 数据字典,Doublewrite Buffer,Change Buffer 的存储区域。系统表空间也默认包含用户在系统表空间创建的表数据和索引数据。系统表空间是一个共享的表空间,因为它是被多个表共享的,该空间的数据文件通过参数 innodb_data_file_path 控制。

默认值为 ibdata1:12M:autoextend(文件名为 ibdata1、12MB、自动扩展)。

2. 独立表空间(File – Per – Tablespaces)

默认开启,独立表空间是一个单表表空间,该表创建于自己的数据文件中,而非

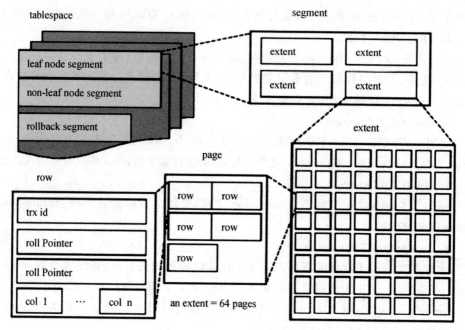

图 6-4 MySQL 逻辑存储架构

创建于系统表空间中。当 innodb_file_per_table 选项开启时,表将被创建于表空间中。否则,innodb 将被创建于系统表空间中。每个文件表空间由一个 .ibd 数据文件代表,该文件默认被创建于数据库目录中。

表空间的表文件支持动态(dynamic)和压缩(commpressed)行格式。

3. 通用表空间

通用表空间为通过 create tablespace 语法创建的共享表空间。通用表空间可以创建于 MySQL 数据目录外的其他位置,其可以容纳多张表,且其支持所有的行格式。

4. 撤销表空间(Undo Tablespaces)

撤销表空间由一个或多个包含 Undo 日志的文件组成,InnoDB 使用的 undo 表空间由 innodb_undo_tablespaces 配置选项控制,默认为 0。

(1)参数值为 0 表示使用系统表空间 ibdata1。

(2)大于 0 表示使用 undo 表空间 undo_001、undo_002 等。

5. 临时表空间(Temporary Tablespaces)

分为 session temporary tablespaces 和 global temporary tablespace 两种。

(1)session temporary tablespaces 存储的是用户创建的临时表和磁盘内部的临时表。

(2)global temporary tablespace 储存用户临时表的回滚段(rollback segments)。

MySQL 服务器正常关闭或异常终止时,临时表空间将被移除,每次启动时会被重新创建。

6.3.2 段、区和行

(1)段。表空间是由段组成的,可以把一个表理解为一个段,通常有数据段、回滚段、索引段等。

区是由连续的页组成的,是物理上连续分配的一段空间,每个区的大小固定是 1 MB。

(2)页。InnoDB 的最小物理存储分配单位是 page,有数据页回滚页等。一般情况下,一个区由 64 个连接的页组成,页默认大小是 16 KB。

(3)行。页里面又记录着行记录的信息,InnoDB 存储引擎是面向行的,也就是数据是按照行存储的,Innodb 存储引擎有常用的文件格式为 Dynamic 和 Compressed 行格式:mysql> show variables like 'innodb%format'.

6.4 表空间运维

6.4.1 表空间创建

1. 通用表空间

(1)//创建表空间 ts1

CREATE TABLESPACE ts1 ADD DATAFILE 'ts1.ibd' Engine = InnoDB

CREATE TABLESPACE 'ts3' ADD DATAFILE '/tmp/ts3.ibd' ENGINE = INNODB

(2)//将表添加到 ts1 表空间

CREATE TABLE t1 (c1 INT PRIMARY KEY) TABLESPACE ts1

(3)//删除表空间,必须删除表

DROP TABLESPACE tablespace_name

(4)//重命名表空间

ALTER TABLESPACE ts3 RENAME TO ts001

2. undo 表空间

CREATE UNDO TABLESPACE undo_003 ADD DATAFILE 'undo_003.ibu'

(1)//停用 UNDO 表空间

ALTER UNDO TABLESPACE undo_003 SET INACTIVE;

(2)//删除 UNDO 表空间

DROP UNDO TABLESPACE undo_003

(3)//启用 UNDO 表空间

ALTER UNDO TABLESPACE undo_003 SET ACTIVE

在数据库初始化的时候就需使用如下三个参数 innodb_undo_tablespaces=3 #设置为 3 个,innodb_undo_directory=/dbfiles/mysql_home/undologs,要在

参数文件中做设置,就可以分离出单独的 undo 表空间。

6.4.2 查看表空间信息

mysql> select * from information_schema.INNODB_TABLESPACES_BRIEF;

mysql> select a.SPACE,a.'NAME',a.SPACE_TYPE,a.ROW_FORMAT,a.STATE

from information_schema.INNODB_TABLESPACES a

6.4.3 查看表空间文件信息

mysql> SELECT FILE_ID,FILE_NAME, FILE_TYPE, TABLESPACE_NAME, STATUS, AUTOEXTEND_SIZE

FROM INFORMATION_SCHEMA.FILES

WHERE TABLESPACE_NAME like 'ts%'

mysql> select file_name,file_type,logfile_group_number,free_extents,total_extents,extra

from INFORMATION_SCHEMA.files where file_type = 'UNDO LOG'

注意:通常,一个 InnoDB 表空间仅支持单个数据文件。如果需要修改 undo 表空间默认路径,需要修改 innodb_undo_directory 变量。如果是创建通用表空间且不在默认路径下,需要配置 innodb_directories 变量。

[AUTOEXTEND_SIZE [=] value] --自动扩展大小

[FILE_BLOCK_SIZE = value] --定义了表空间数据文件的块大小。

6.5 分区表

6.5.1 表分区的概念

我们可以通过 show variables like'%datadir%'命令来查看数据文件存放的默认路径,一个数据库就是一个文件夹,一个库中。只要一张表的数据量过大,就会导致 *.ibd 文件过大,数据的查找就会变得很慢。

MySQL 从 5.1 版本开始添加了对分区的支持,分区的过程是将一个表或索引分解为多个更小、更可管理的部分。对于开发者而言,分区后的表使用方式和不分区基本上是一样的,只不过在物理存储上,该表原本只有一个数据文件,现在变成了多个,每个分区都是独立的对象,可以独自处理,也可以作为一个更大的对象一部分进行处理,表分区如图 6-5 所示。

常见的存储引擎如 InnoDB、MyISAM、NDB 等都支持分区。但并不是所有的存储引擎都支持,如 CSV、FEDORATED、MERGE 等就不支持分区,因此在使用此分

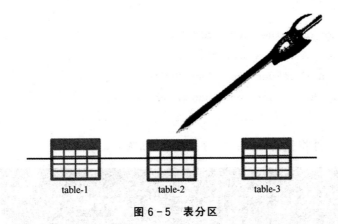

图 6-5 表分区

区功能前,应该对选择的存储引擎分区支持有所了解。

6.5.2 进行表分区的理由

1. 分区的优点

(1)可以让单表存储更多的数据。

(2)分区表的数据更容易维护,可以通过清除整个分区批量删除大量数据,也可以增加新的分区来支持新插入的数据。另外,还可以对一个独立分区进行优化、检查、修复等操作。

(3)部分查询能够从查询条件确定只落在少数分区上,查询速度会很快。

(4)分区表的数据还可以分布在不同的物理设备上,从而高效利用多个硬件设备。

(5)可以使用分区表来避免某些特殊瓶颈,例如 InnoDB 单个索引的互斥访问、ext3 文件系统的 inode 锁竞争。

(6)可以备份和恢复单个分区。

2. 分区的缺点

(1)一个表最多只能有 1 024 个分区。

(2)如果分区字段中有主键或者唯一索引的列,那么所有主键列和唯一索引列都必须包含进来。

(3)分区表无法使用外键约束。

(4)NULL 值会使分区过滤无效。

(5)所有分区必须使用相同的存储引擎。

6.5.3 分区实践

1. 创建分区表

create table user(id int(11) not null,name varchar(32) not null)
partition by range(id)

(
   partition p0 values less than(10)
   partition p1 values less than(20)
   partition p2 values less than(30)
   partition p3 values less than maxvalue
)

数据存储文件将根据分区被拆分成多份，如图 6-6 所示。

```
-rw-r----- 1 mysql mysql 114688 Apr  4 23:24 d1.ibd
-rw-r----- 1 mysql mysql 114688 Apr  3 18:00 d2.ibd
-rw-r----- 1 mysql mysql 114688 Apr  3 20:18 test_collate.ibd
-rw-r----- 1 mysql mysql 114688 Apr  9 14:55 user#p#p0.ibd
-rw-r----- 1 mysql mysql 114688 Apr  9 14:55 user#p#p1.ibd
-rw-r----- 1 mysql mysql 114688 Apr  9 14:55 user#p#p2.ibd
-rw-r----- 1 mysql mysql 114688 Apr  9 14:55 user#p#p3.ibd
```

图 6-6　表分区存储文件

2. 分区表查询

```
insert into user values(1,'IT')
insert into user values(12,'007')
insert into user values(22,'jeames')
insert into user values(50,'TenKE')

select * from user partition(p0)
select * from user partition(p1)
select * from user partition(p2)
select * from user partition(p3)
```

新增几条数据后查询可以看到数据已经分散在不同的分区中，如图 6-7 所示。

```
mysql> select * from user partition(p0);
+----+------+
| id | name |
+----+------+
|  1 | IT   |
+----+------+
1 row in set (0.00 sec)
```

图 6-7　表分区数据分散

```
mysql> select * from information_schema.partitions
mysql> select * from information_schema.partitions
where table_schema = 'jeames' and table_name = 'user'\G
```

6.5.4 表分区的类型

1. RANGE 表分区

RANGE 表分区,即按照一定的范围值来确定每个分区包含的数据,如使用 range 表分区创建语句是:partition by range(id) partition p0 values less than()。

分区的定义范围必须是连续的,且不能重叠,使用 values less than()来定义分区范围,且用从小到大的顺序定义范围。给分区字段赋值的时候,分区字段取值范围不能超过 values less than()的取值范围。使用 values less than maxvalue 来将未来不确定的值放到这个表分区中。

按时间类型(datetime)来做表分区可以在 RANGE()中使用函数来做转换。

例:partition by range(year(create_time)),timestamp 可以使用 unix_timestamp('2019-11-20 00:00:00')转化。

```
create table user_range(
  id int
  username varchar(255)
  password varchar(255)
  createDate date
  primary key (id,createDate)
)engine= innodb
  partition by range(year(createDate))(
      partition p2022 values less than(2023)
      partition p2023 values less than(2024)
      partition p2024 values less than(2025)
)
```

注意:createDate 是联合主键的一员,如果 createDate 不是主键,只是一个普通字段,那么创建时就会抛出错误:ERROR 1503 的报错。

```
--删除分区
alter table user_range drop partition p2022

--新增分区
alter table user_range add partition(partition p2025 values less than
(2026))
```

2. LIST 表分区

列表分区,如果按照一个一个确定的值来确定每个分区包含的数据,创建语句是:

```
partition by list(id)    partition p0 values in(1,2,3)
```

分区字段必须是整数类型或者分区函数返回整数,取值范围通过 values in( )来定义,不能使用 maxvalue。

假设有一个用户表,用户有性别,现在想按照性别将用户分开存储,男性存储在一个分区中,女性存储在一个分区中,SQL 如下:

```
create   table user_list(
  id int
  username varchar(255)
  password varchar(255)
  gender int
  primary key(id, gender)
)engine= innodb
  partition by list(gender)(
    partition  man  values  in  (1)
    partition  woman  values  in  (0)
  )
Insert into user_list    values(1,'zhangsan','123',1)
Insert into user_list    values(2,'wangwu','123',3)
```

3. HASH 表分区

哈希表分区,按照一个自定义的函数返回值来确定每个分区包含的数据,HASH 分区的目的是将数据均匀地分布到预先定义的各个分区中,保证各分区的数据量大致都是一样的。在 RANGE 和 LIST 分区中,必须明确指定一个给定的列值或列值集合应该保存在哪个分区中;而在 HASH 分区中,MySQL 会自动完成这些工作,用户所要做的只是基于将要进行哈希分区的列指定一个表达式,及分区的数量。

使用 HASH 分区来分割一个表,要在 CREATE TABLE 语句上添加 PARTITION BY HASH（expr),其中 expr 是一个字段或是一个返回整数的表达式;另外通过 PARTITIONS 属性指定分区的数量,如果没有指定,那么分区的数量默认为 1,另外,HASH 分区不能删除分区,所以不能使用 DROP PARTITION 操作进行分区删除操作。

```
create   table user_hash(
  id int
  username varchar(255)
  password varchar(255)
```

```
    gender int
    primary key(id, gender)
)engine= innodb partition by hash(id) partitions 4
```

注意：partition by hash(id) partitions 4。

根据 hash 算法来分配到分区中，以上设置四个分区，并根据 id%4 进行取模运算，根据余数插入到指定的分区中。

4. KEY 分区

KEY 分区和 HASH 分区相似，但是 KEY 分区支持除 text 和 BLOB 之外的所有数据类型的分区，而 HASH 分区只支持数字分区。KEY 分区不允许使用用户自定义的表达式进行分区，KEY 分区使用系统提供的 HASH 函数进行分区。当表中存在主键或者唯一索引时，如果创建 KEY 分区时没有指定字段系统默认，则会首选主键列作为分区字段，如果不存在主键列，会选择非空唯一索引列作为分区字段。

key()括号里面可以包含 0 个或多个字段（不必是整数类型，可以是普通字段）。

```
create table user_key(
    id int
    username varchar(255)
    password varchar(255)
    gender int
    primary key(id, gender)
)engine= innodb partition by key() partitions 4
```

5. range、list 分区

range、list 分区可以指定多个字段作为分区字段，而 COLUMN 分区是 5.5 版本开始引入的分区功能，只有 RANGE COLUMN 和 LIST COLUMN 两种分区，且其支持整形、日期、字符串，这种分区方式和 RANGE、LIST 的分区方式非常相似。针对日期字段的分区不需要再使用函数进行转换了。

COLUMN 分区支持多个字段作为分区键但是不支持表达式作为分区键。
COLUMNS 支持的类型有：
（1）整形支持：tinyint、smallint、mediumint、int、bigint；不支持 decimal 和 float。
（2）时间类型支持：date、datetime。
（3）字符类型支持：char、varchar、binary、varbinary；不支持 text、blob。

多字段分区创建示例：
```
create table user1(
    id int
    username varchar(255)
```

```
    password varchar(255)
    gender int
    createDate date
    primary key(id, createDate)
)engine= innodb PARTITION BY RANGE COLUMNS(createDate) (
    PARTITION p0 VALUES LESS THAN ('1990-01-01')
    PARTITION p1 VALUES LESS THAN ('2000-01-01')
    PARTITION p2 VALUES LESS THAN ('2010-01-01')
    PARTITION p3 VALUES LESS THAN ('2020-01-01')
    PARTITION p4 VALUES LESS THAN MAXVALUE
)
create table user2(
    id int
    username varchar(255)
    password varchar(255)
    gender int
    createDate date
    primary key(id, createDate)
)engine= innodb PARTITION BY LIST COLUMNS(createDate) (
    PARTITION p0 VALUES IN ('1990-01-01')
    PARTITION p1 VALUES IN ('2000-01-01')
    PARTITION p2 VALUES IN ('2010-01-01')
    PARTITION p3 VALUES IN ('2020-01-01')
)
```

6.5.5 分区表运维命令

1. 添加分区

alter table user add partition (partition p3 values less than (4000)); -- range 分区

alter table user add partition (partition p3 values in (40)); -- lists 分区

2. 删除表分区(会删除数据)

alter table user drop partition p30

3. 删除表的所有分区(不会丢失数据)

alter table user_list remove partitioning

4. 重新定义 range 分区表(不会丢失数据)

alter table user_list partition by list(gender)

(
partition man values in (1)
partition woman values in (0)
)

5. 重新定义 hash 分区表(不会丢失数据)

alter table user partition by hash(salary) partitions 7

6. 合并分区:把 2 个分区合并为一个,不会丢失数据

alter table user reorganize partition p1,p2 into (partition p1 values less than (1000))

6.6 线程结构

6.6.1 后台线程

MySQL 是一个单进程多线程架构的数据库管理系统,所以当 MySQL 实例启动之后,会存在众多的线程来做各种各样的事情。我们可以通过 performance_schema.threads 表查询到所有线程,包括后台线程和前台线程,如图 6-8 所示。

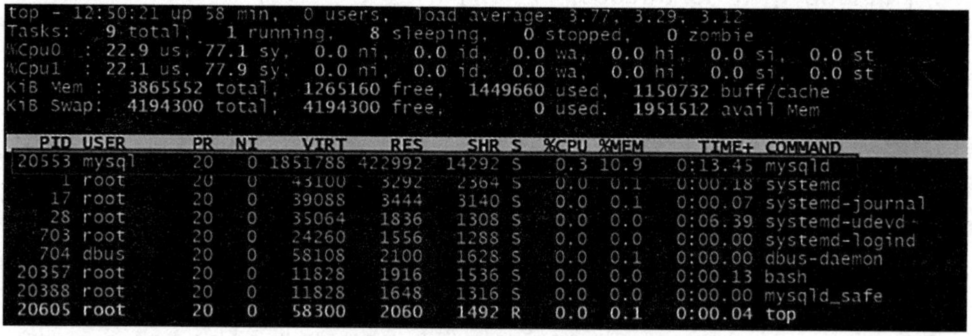

图 6-8 MySQL 单进程多线程

表查询到所有线程,包括后台线程和前台线程,这里主要介绍后台线程。

mysql> select name,count(*) from performance_schema.threads where type = 'BACKGROUND' group by name

mysql> select name , type , thread_id,processlist_id from performance_schema.threads where type = 'BACKGROUND'

这些后台线程的主要功能如下:

（1）srv_master_thread（主线程）：InnoDB 存储引擎主线程，由 4 个循环组成，即主循环(loop)、后台循环(background loop)、刷新循环(flush loop)、暂停循环(suspend loop)，大多数工作都在主循环中完成，主循环主要负责将脏缓存页刷新到数据文件中，执行 undo purge 操作，触发检查点，合并插入缓冲区等。

（2）io_ibuf_thread（插入缓冲线程）：主要负责插入缓冲区的合并操作。将对辅助索引页的修改操作从随机变成顺序 I/O，大幅度提升了效率（会先判断发生修改的辅助索引页是否在缓冲池中，如果在则直接修改；如果不在，则先存放在一个 Change Buffer 对象中。当其读取操作把该修改对应的页从磁盘读取到缓冲池中时，就会合并该 Change Buffer 对象中保存的记录到辅助索引页中)。

（3）io_read_thread（读 I/O 操作线程）：负责数据库的 AIO（异步 I/O）读取操作，可以配置多个读线程，由参数 innodb_read_io_threads 设置，默认值为 4。

（4）io_write_thread（写 I/O 操作线程）：负责数据库的 AIO（异步 I/O）写操作，可配置多个写线程，由参数 innodb_write_io_threads 设置，默认值为 4。

```
mysql> show variables like '%io_threads%'
+------------------------+-------+
| Variable_name          | Value |
+------------------------+-------+
| innodb_read_io_threads  | 4     |
| innodb_write_io_threads | 4     |
+------------------------+-------+
```

（5）io_log_thread（日志线程）：用于将重做日志刷新到日志文件中。

（6）srv_purge_thread（undo 清理线程）：主要负责 undo 页清理操作，在 MySQL 5.6 之后可设置独立的线程执行 undo purge 操作，以减少主线程负载。通过 innodb_purge_threads 参数来控制 purge 的线程个数，默认是 1 个。

（7）srv_lock_timeout_thread（锁线程）：负责锁控制和死锁检测等。

（8）srv_error_monitor_thread（错误监控线程）：主要负责错误控制和错误处理。

（9）page_cleaner_thread（脏页清理线程）：负责刷脏操作，在 MySQL 5.6 之后可设置独立的线程执行刷脏操作，以减少主线程负载。刷脏分为两种方式，其中一种是基于 LRU list（最近最少访问的列表）的最近访问时间顺序刷新的；另一种是基于 flush list（刷新列表）的最近修改时间和 LSN 值的顺序刷新的。

（10）thread_timer_notifier（计时器过期通知线程）：超过计时器时间，通知线程处理。一般在超过 MAX_EXECUTION_TIME 时自动中止 SQL 语句的执行。

（11）main（主线程）：MySQL 服务的主线程（请与 InnoDB 存储引擎主线程区别开)，包括初始化、读取配置文件等功能。

（12）srv_monitor_thread（InnoDB 监控打印线程）：如果开启了 InnoDB 监控器，

那么每隔 5 秒打印一次 InnoDB 监视器采集的信息。

(13) srv_worker_thread(InnoDB 工作线程)：InnoDB 的实际工作线程，轮询从任务队列中取出任务(row select、row insert 等)并执行。

(14) buf_dump_thread(InnoDB 缓冲池导入/导出线程)：InnoDB 缓冲池热点数据页导入/导出的线程。

(15) dict_stats_thread(InnoDB 后台统计线程)：负责 InnoDB 表统计信息更新的后台线程。

(16) signal_handler(信号处理线程)：负责 SIGTERM、SIGQUIT、SIGHUP 信号处理的线程。

6.6.2 前台线程

前台线程其实就是通过 TCP/IP 协议或客户端程序创建的线程，所以包括了与主备复制相关的线程和用户创建的连接线程。跟 MySQL 后台线程类似，我们也可以通过 performance_schema.threads 表来查询 MySQL 中有哪些前台线程。

```
mysql> select * from performance_schema.threads
mysql> select name,type,thread_id,processlist_id from performance_schema.threads where type='FOREGROUND'
mysql> show processlist
```

可以看到，有 5 种前台线程，这些前台线程的主要功能如下。

(1) compress_gtid_table(GTID 压缩线程)：用于压缩 MySQL 5.7 新增的 mysql.gtid_executed 表中的 GTID 记录数量。在 MySQL 5.7 版本中，当从库关闭 log-bin 或者 log_slave_updates 参数之后，SQL 线程每应用一个事务就会实时更新一次 mysql.gtid_executed 表(在 MySQL 5.7 中启用 GTID 复制时可以关闭 log_slave_updates 参数，使用该表来记录 GTID。但在 MySQL 5.6 中由于没有此表，所以不能关闭 log_slave_updates 参数)，时间一长，该表中就会存在大量的 GTID 记录(每个事务一行)，使用该线程可以把多行记录压缩成一行。

(2) one_connection(用户连接线程)：用于处理用户请求的线程。

(3) slave_io(I/O 线程)：用于拉取主库 binlog 日志的线程。

(4) slave_sql(SQL 线程)：用于应用从主库拉取的 binlog 日志的线程。

注意：在多线程复制中，该线程为协调器线程，用于分发 binlog 日志给工作线程(slave_worker)应用，并对多个工作线程进行协调。

(5) slave_worker(工作线程)：在多线程复制场景中，接收并应用 SQL 线程(slave_sql)分发的主库 binlog 日志，多个工作线程之间的一致性依靠 SQL 线程(slave_sql)进行协调。

Part 4 备份恢复篇

第 7 章 逻辑备份

7.1 MySQL 导入导出

MySQL 作为一个数据库系统,其备份自然也是非常重要且有必要去做。备份的理由千千万,预防故障,安全需求,回滚,审计,删了又改的需求等,备份的重要性不言而喻。MySQL 支持的备份类型如图 7-1 所示。

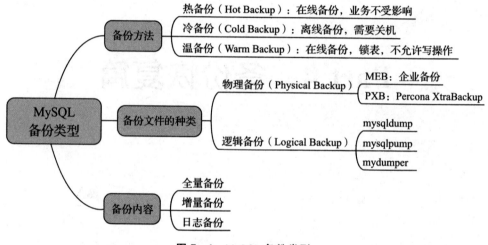

图 7-1 MySQL 备份类型

使用 cp 进行冷备份案例:

(1)关闭 MySQL 数据库服务器

(2)拷贝 MySQL 数据库的数据文件

mkdir /backup #创建文件夹存放备份数据库文件

cp -a /var/lib/mysql/* /backup #保留权限的拷贝源数据文件

ls /backup #查看目录下的文件

(3)恢复数据库

#将备份的数据文件拷贝回去

cp -a /backup/* /var/lib/mysql/

#重启 MySQL

systemctl start mysqld

日常的运维中我们常用 SELECT…INTO OUTFILE、mysqldump、mysql-e 等导出数据,同时用 LOAD DATA INFILE、MySQLimport 等导入数据,另外一些客户端工具,如 navicat、dbeaver、workbench 等也支持导入导出。

7.1.1 SELECT…INTO OUTFILE

SELECT INTO … OUTFILE 语句把表数据导出到一个文本文件中,并用 LOAD DATA …INFILE 语句恢复数据。但是这种方法只能导出或导入数据的内容,不包括表的结构,如果表的结构文件损坏,则必须先恢复原来的表的结构,导出的文件只能在服务器上。

(1)SELECT ... INTO OUTFILE 是将表中的数据写到一个文件。

(2)LOAD DATA INFILE 则是将文件内容导入表。

语法结构:
SELECT
[INTO OUTFILE 'file_name'
　　　[CHARACTER SET charset_name]
　　　export_options
  | INTO DUMPFILE 'file_name'
  | INTO var_name [, var_name]]

语法解析:

(1) CHARACTER SET:数据被转换成 CHARACTER SET 指定的编码格式输出。如果不指定的话默认为 binary,即不做转换,如果指定多个编码格式,输出文件将不能被正确载入。

(2) export_options:用于语句的 exort_options 部分的语法包括部分 FIELDS 和 LINES 子句,这些子句与 LOAD DATA INFILE 语句同时使用。

以上的导入导出,有个非常重要的参数 secure-file-priv,详细说明如下:

secure-file-priv 参数是用来限制 LOAD DATA,SELECT ... OUTFILE 及 LOAD_FILE()传到哪个指定目录的。

①seure_file_priv 的值为 null ,表示限制 mysqld 不允许导入|导出。

②当 secure_file_priv 的值为/etc/ ,表示限制 mysqld 的导入|导出只能发生在/etc/目录下。

③当 secure_file_priv 的值没有具体值时,表示不对 mysqld 的导入|导出做限制。

修改参数只能在服务器参数文件中修改,修改后重新服务器:

secure-file-priv=

注意: 8.0 版本需要设置参数 secure-file-priv 为非 NULL,必须设置

```
mysql> show global variables like '%secure%';
+-------------------------+--------------------+
| Variable_name           | Value              |
+-------------------------+--------------------+
| require_secure_transport | OFF               |
| secure_file_priv        | /var/lib/mysql-files/ |
+-------------------------+--------------------+
```

(3) SELECT…INTO OUTFILE 导出

```
select host,user from mysql.user into outfile '/tmp/user.csv'
Fields terminated by ',' enclosed by '"';
```

说明：逗号分割，双引号闭合。

(4) LOAD DATA INFILE 导入

```
mysql> show variables like '%local%'
+---------------+-------+
| Variable_name | Value |
+---------------+-------+
| local_infile  | OFF   |
+---------------+-------+

mysql> set global local_infile = 1;
--此处一定要加字段分隔符号,注意格式。

mysql> LOAD DATA LOCAL INFILE 'e:\user.csv' INTO TABLE db1.user FIELDS TERMINATED BY ','
```

注意：取消 LOCAL 则表示从服务器指定路径导入。

7.1.2 mysqldump

mysqldump 是一个数据导出工具,能够导出数据库中的数据并进行备份,或将数据迁移到其他数据库中,mysqldump 命令的使用格式如下：

```
shell> mysqldump [options] db_name [tbl_name ...]
shell> mysqldump [options] --databases db_name ...
shell> mysqldump [options] --all-databases
```

其中,options 为 mysqldump 命令的一些选项,可以使用如下命令查阅。

```
mysqldump --help
```

(1) mysqldump 命令导出

```
[root@jeames ~ ]#   mkdir /dumpbak
[root@jeames ~ ]#   chown mysql:mysql /dumpbak
```
##远程
```
mysqldump-uroot-proot-h192.168.1.88-P3306 db1-T /dumpbak
```
##本地服务器
```
mysqldump-uroot-proot test-T /dumpbak
```
说明：

-T 后面必须加目录，自动生成两个文件：

一个.sql 文件，创建表结构的语句

一个.txt 文件，数据文件

(2)将数据库 db_name 中的 t1 及 t2 表导出到 dumpbak 目录中，多个表中间使用空格

```
mysqldump-uroot-proot-h192.168.66.35-P3317db_name--tables t1 t2-T /dumpbak
```

(3)将数据库 mes_db 导出到目录/bk 中，字段使用双引号用逗号隔开

```
mysqldump-uroot-proot--fields-terminated-by ','--fields-enclosed-by '"' mes_db-T /bk
```

①mysqldump 和 mysqld 在同一个主机上，可以生成.sql 和.txt 文件

②mysqldump 和 mysqld 在不同主机上，只生成.txt 文件

③mysqldump 和 mysqld 不能位于不同操作系统，否则会报

③如果在 Linux 系统，可以使用如下命令转换格式

```
find ./-name "*.txt" | awk -F "." '{print $2}' | xargs -i-t mv ./{}.txt ./{}.csv
```

7.1.3 MySQLimport

mysqlimport 命令主要用来向数据库中导入数据，如果使用 mysqldump 命令导出数据时使用了-T 参数，则可以使用 mysqlimport 命令将 mysqldump 导出的文件内容导入数据库中。

(1)mysqlimport 命令的使用格式

shell>   mysqlimport [options] db_name textfile1 [textfile2 ...]

其中,关于 options 的具体信息,可以使用如下命令查阅。

```
mysqlimport --help
```
(2)MySQLimport 命令导入

①将数据库 mes_db 中的表 student 导出到目录/bk 中,字段使用双引号用逗号

隔开

mysqldump -uroot -proot --fields-terminated-by ',' --fields-enclosed-by '"' mes_db student /bk/student.csv

②将本地的 student.csv 导入到另外一个 MySQL 数据库 db1 中

mysql> set global local_infile = 1

mysql> mysqlimport -uroot -proot -h192.168.1.54 -P3310 --local db1 "e:\student.csv" --fields-terminated-by= ","

7.2 Mysqldump 逻辑备份恢复

Mysqldump 是 mysql 自带的备份工具，一般在如下目录中：
/usr/local/mysql/bin/mysqldump，支持基于 innodb 的热备份。但是由于是逻辑备份，所以速度不是很快，适合备份数据比较小的场景，是一个客户端工具。

实际应用中，一般都是基于 Mysqldump 完全备份＋二进制日志可以实现基于时间点的恢复。

7.2.1 Mysqldump 导出

Mysqldump 客户机实用程序执行逻辑备份，生成一组 SQL 语句，可以执行这些语句来重现原始数据库对象定义和表数据。它将一个或多个 MySQL 数据库转储以备份或传输到另一个 SQL server。Mysqldump 命令还可以生成 CSV、其他分隔文本或 XML 格式的输出。

通常有三种方法可以使用 Mysqldump 转储一组或多个表、一组或更多完整数据库或整个 MySQL 服务器，如下所示：

mysqldump [options] db_name [tbl_name ...]

mysqldump [options] --databases db_name ...

mysqldump [options] --all-databases

Mysqldump 导出案例：

(1) 导出单库

mysqldump -uroot -proot -h192.168.1.88 -P3306 test > e:\test.sql

(2) 导出单表

mysqldump -uroot -proot -h192.168.1.54 -P3310 db1 student> e:\db1_student.sql

(3) 导出多个库

mysqldump -uroot -proot -h192.168.1.54 -P3310 --databases db1 mysql> e:\db1_mysql.sql

(4)导出所有数据库

mysqldump‐uroot‐proot‐h192.168.1.54‐P3310‐A> e:\all.sql

mysqldump‐uroot‐proot‐h192.168.1.54‐P3310‐‐all‐databases> e:\all.sql

(5)Linux 服务器可做如下操作

mysqldump‐uroot‐proot‐‐all‐databases> /dumpbak/all.sql

7.2.2 Mysqldump 恢复

恢复的方法：mysql> source d:\script.sql

或者在 dump 文件路径下执行 mysql‐uroot‐proot< data.sql

【案例1】

MySQL 全库备份,如何只恢复一个库？

(1)备份全库

mysqldump‐uroot‐proot‐‐all‐databases> /bk/all.sql

(2)指定当前数据库为 db_school

sed‐n '/^‐‐ Current Database:'db_school'/,/^‐‐ Current Database: '/p' all.sql> db1.sql

(3)导入

mysql‐uroot‐pxx‐D db1< db1.sql

【案例2】

MySQL 全库备份,如何只恢复单个表的数据

(1)导出建表语句,test 为表,登录数据库建表

sed‐e '/./{H;$!d;}'‐e 'x;/CREATE TABLE 'test'/!d;q' all.sql > ddl.sql

(2)生成插入语句

grep‐i 'INSERT INTO 'test'' all.sql> data.sql

(3)导入数据

mysql‐uroot‐proot‐D db1 < data.sql

7.2.3 Mysqldump 迁移数据库

1. 在线导出数据库

mysqldump‐uroot‐proot‐h192.168.1.54‐P3310‐‐single‐transaction‐‐

hex-blob --routines --events --triggers --master-data=2 --databases db_school --default-character-set=utf8 --max_allowed_packet=512M > e:\db_school.sql

参数：

\#--single-transaction 表示一致性备份

\#--hex-blob 表示可以导出二进制数据，包括图片或者文字，blob

\#--routines 表示存储过程和函数

\#--events 表示 job

\#--triggers 表示触发器

\#--master-data=2 表示 dump 文件中包含了创建主从复制的相关 sql

\#max_allowed_packet 是 Mysql 中的一个设定参数，用于设定所接受的包的大小

导出参数设置（根据实际业务量设定）：

set global wait_timeout=28800000
set global net_read_timeout=28800
set global net_write_timeout=28800
set global max_allowed_packet=1*1024*1024*1024

2. 目标库导入

mysql-uroot-proot-f --default-character-set=utf8 < db_school.sql

导入时设置（加快导入的速度）

set sql_log_bin=0
set global innodb_flush_log_at_trx_commit = 2
set global sync_binlog = 20000
set global max_allowed_packet=1*1024*1024*1024
set global net_buffer_length=100000
set global interactive_timeout=28800000
set global wait_timeout=28800000

注意：

(1)如果单独导出表，表上有触发器，则会自动导出触发器。

(2)如果用的是 Windows 下的 cmd 中导入，则需要根据导出的 sql 文件字符集来设置代码页（默认为 UTF8），否则可能会报莫名其妙的错误，ERROR at line 3759：Unknown。

(3)导出导入字符设置参数：--default-character-set＝utf8。

3. 迁移后确认

确认表、事件、触发器、存储过程、函数等数量是否吻合。

select db 数据库,type 对象类型,cnt 对象数量 from
(select '表' type, table_schema db, count(*) cnt from information_schema.'TABLES' a where table_type='BASE TABLE' group by table_schema
union all
select '事件' type,event_schema db,count(*) cnt from information_schema.'EVENTS' b group by event_schema
union all
select '触发器' type,trigger_schema db,count(*) cnt from information_schema.'TRIGGERS' c group by trigger_schema
union all
select '存储过程' type,routine_schema db,count(*) cnt from information_schema.ROUTINES d where'ROUTINE_TYPE' = 'PROCEDURE' group by db
union all
select '函数' type,routine_schema db,count(*) cnt from information_schema.ROUTINES d where'ROUTINE_TYPE' = 'FUNCTION' group by db
union all
select '视图' type,table_schema db,count(*) cnt from information_schema.VIEWS f group by table_schema) t
where db= 'db_school'
order by db,type;

7.3 Mysqlpump

MySQL5.7 之后多了一个备份工具：Mysqlpump。它是 Mysqldump 的一个衍生，Mysqlpump 和 Mysqldump 一样，属于逻辑备份,备份以 SQL 形式的文本保存。逻辑备份相对物理备份的好处是不关心 undo log 的大小,直接备份数据即可,默认情况下仅备份：表,视图和触发器,应用时建议使用,它最主要的特点是：

(1)并行备份数据库和数据库中的对象的,加快备份过程。

(2)更好控制数据库和数据库对象(表,存储过程,用户帐户)的备份。

(3)备份用户账号作为账户管理语句(CREATE USER,GRANT),而不是直接插入到 MySQL 的系统数据库。

(4)备份出来直接生成压缩后的备份文件。

(5)备份进度指示(估计值)。

(6)重新加载(还原)备份文件,先建表后插入数据最后建立索引,减少了索引维护开销,加快了还原速度。

(7)备份可以排除或指定数据库。

重点参数：

(1)支持基于表的多线程导出功能(-- default - parallelism,默认为 2,-- parallel - schemas,控制并行导出的库)。

(2)导出的时候带有进度条(-- watch - progress,默认开启)。

(3)支持直接压缩导出导入(参数-- compress - output,而且支持 ZLIB 和 LZ4)。

Mysqlpump 导出案例：

(1)备份 db_school 库下的 student 表

mysqlpump - uroot - proot - h192. 168. 1. 54 - P3310 db_school student -- single - transaction -- set - gtid - purged= OFF -- default - parallelism= 4 > /dumpbak/db_school_student. sql

(2)备份 db_school 库

mysqlpump - uroot - proot db_school -- single - transaction -- default - parallelism= 4 > /dumpbak/db_school. sql

(3)备份 db_school 库和 db1 库

mysqlpump - uroot - proot sbtest -- databases db_school db1 -- single - transaction -- default - parallelism= 4 > /dumpbak/db_school_db1. sql

(4)备份所有数据库

mysqlpump - uroot - proot -- all - databases -- single - transaction -- default - parallelism= 4 > /dumpbak/all_fullbak. sql

注意：多个表、多个 DB 可以并行导出(多线程并行导出),单个表不支持并行。

第8章 物理备份

8.1 PXB 介绍及部署

8.1.1 PXB 介绍

PXB(Percona XtraBackup)是由 Percona 公司提供的一款 MySQL 数据库备份恢复工具,是一款开源的能够对 Innodb 和 xtradb 存储引擎数据库进行热备的工具(备份时不影响数据读写)。

MySQL 冷备、mysqldump、MySQL 热拷贝都无法实现对数据库进行增量备份,在生产环境中增量备份是非常实用的,如果数据不到 1T 的时候,存储空间足够的情况下,可以每天进行完整备份,如果每天产生的数据量较大,需要定制数据备份策略。例如每周实用完整备份,周一到周六实用增量备份,而 Percona - Xtrabackup 就是为了实现增量备份而出现的一款主流备份工具。它有如下特点:

(1)备份过程快速、可靠。
(2)备份过程不会打断正在执行的事务。
(3)能够基于压缩等功能节约磁盘空间和流量。
(4)自动实现备份检验。
(5)还原速度快。

8.1.2 PXB 部署

官网:https://www.percona.com/。
安装包下载地址:https://www.percona.com/downloads/。
注意事项:
安装 PXB 的版本一定 mysql 的版本兼容,比如 8.0 版本要保持一致
PXB 的版本 2.4 只支持 mysql5.6、5.7。
PXB 一定要和 mysql 服务安装在一起。
一般,公司服务器为了安全期间,不支持访问外网,故采用二进制方式安装 8.0 版本。
二进制包位置下载:https://github.com/percona。
1. 安装依赖包

yum - y install perl perl - devel libaio libaio - devel perl - Time - HiRes perl-DBD-MySQL numactl

2. 解压缩安装包

[root@jeames ~]# tar-zxvf percona-xtrabackup-8.0.14-Linux-x86_64.glibc2.12.tar.gz

3. 软连接设置

[root@jeames ~]# mv percona-xtrabackup-8.0.14-Linux-x86_64.glibc2.12 \

/usr/local/percona-xtrabackup-8.0.14-Linux-x86_64

[root@jeames ~]# ln -s /usr/local/percona-xtrabackup-8.0.14-Linux-x86_64 \ /usr/local/xtrabackup-8.0.14

[root@jeames ~]# ln -s /usr/local/percona-xtrabackup-8.0.14-Linux-x86_64/bin/xtrabackup \ /usr/bin/xtrabackup8014

[root@jeames ~]# ln -s /usr/local/percona-xtrabackup-8.0.14-Linux-x86_64/bin/xbstream \

/usr/bin/xbstream8014

[root@jeames ~]# which xtrabackup8014

[root@jeames ~]# xtrabackup8014 --help

8.2 PXB 备份恢复

Percona Xtrabackup 8.0 版本只能备份 8.0 版本的 MySQL，使用 xtrabackup 命令 2.4 版本支持 MySQL 5.11，5.5，5.6 和 5.7 的版本。

8.2.1 全量备份和恢复

1. 备份

[root@jeames ~]# mkdir -p /bk/

[root@jeames ~]# xtrabackup -uroot -proot -S/var/lib/mysql/mysql.sock \

--backup --target-dir=/bk/full

详细的备份过程如图 8-1 所示。
备份文件：
（1）xtrabackup_binlog_info：服务器当前正在使用的二进制日志文件及至备份这一刻为止二进制日志事件的位置。
（2）xtrabackup_checkpoints：备份类型（如完全或增量）、备份状态（如是否已经

```
[root@pxb /]# xtrabackup -uroot -proot -S/var/lib/mysql/mysql.sock --backup --target-dir=/bk/full
xtrabackup: recognized server arguments: --datadir=/var/lib/mysql
xtrabackup: recognized client arguments: --user=root --password=* --socket=/var/lib/mysql/mysql.sock --backup=1 --target-dir=/bk/full
xtrabackup version 8.0.27-19 based on MySQL server 8.0.27 Linux (x86_64) (revision id: 50dbc8dadda)
perl: warning: Setting locale failed.
perl: warning: Please check that your locale settings:
        LANGUAGE = (unset),
        LC_ALL = (unset),
        LANG = "en_US.UTF-8"
    are supported and installed on your system.
perl: warning: Falling back to the standard locale ("C").
220415 15:13:42  version_check Connecting to MySQL server with DSN 'dbi:mysql:;mysql_read_default_group=xtrabackup;mysql_socket=/var/lib/mysql/mysql.sock' as 'root'  (using password: YES).
220415 15:13:42  version_check Connected to MySQL server
220415 15:13:42  version_check Executing a version check against the server...
220415 15:13:42  version_check Done.
220415 15:13:42 Connecting to MySQL server host: localhost, user: root, password: set, port: not set, socket: /var/lib/mysql/mys
```

图 8-1 PXB 全量备份

为 prepared 状态)和 LSN(日志序列号)的范围信息。

每个 InnoDB 页(通常为 16k 大小)都会包含一个日志序列号,即 LSN,LSN 是整个数据库系统的系统版本号,每个页面相关的 LSN 能够表明此页面最近是如何发生改变的。

(3)xtrabackup_info:备份软件的信息。

(4)backup-my.cnf:备份命令用到的配置选项信息

2. 恢复

恢复方法采用 prepare + copy-back

[root@jeames ~]# xtrabackup -- prepare -- target-dir= /bk/full

此时关闭数据库 systemctl stop mysqld

[root@jeames ~]# xtrabackup -- copy - back -- target - dir= /bk/full -- datadir= /var/lib/mysql

[root@jeames ~]# chown - R mysql:mysql /var/lib/mysql

(1)prepare 的作用 一般情况下,在备份完成后,数据尚且不能用于恢复操作,因为备份的数据中可能会包含尚未提交的事务或已经提交但尚未同步至数据文件中的事务。因此,此时数据文件仍处于不一致状态。"准备"的主要作用正是通过回滚未提交的事务及同步已经提交的事务至数据文件使得数据文件处于一致性状态。在实现"准备"的过程中,xtrabackup 通常还可以使用-- use - memory 选项来指定其可以使用的内存的大小,默认通常为 100M。如果有足够的内存可用,可以多划分一些内存给 prepare 的过程,以提高其完成速度。

(2)xtrabackup 命令的-- copy - back 选项用于执行恢复操作,其通过复制所有数据相关的文件至 mysql 服务器 DATADIR(数据文件)目录中来执行恢复过程。

(3)当数据恢复至 DATADIR 目录以后,还需要确保所有数据文件的属主和属组均为正确的用户,如 mysql,否则,在启动 mysqld 之前还需要事先修改数据文件的属主和属组。

8.2.2 增量备份和恢复

需要注意的是，增量备份仅能应用于 InnoDB 或 XtraDB 表，对于 MyISAM 表而言，执行增量备份时其实进行的是完全备份。"准备"（prepare）增量备份与整理完全备份有着一些不同，尤其要注意的是：

（1）需要在每个备份（包括完全和各个增量备份）上，将已经提交的事务进行"重放"。"重放"之后，所有的备份数据将合并到完全备份上。

（2）基于所有的备份将未提交的事务进行"回滚"。

在增量备份命令中，--incremental-basedir 指的是上一次的增量备份所在的目录。若是一级增量备份，则指向全备的目录。

增量备份策略如下图 8-2 所示。

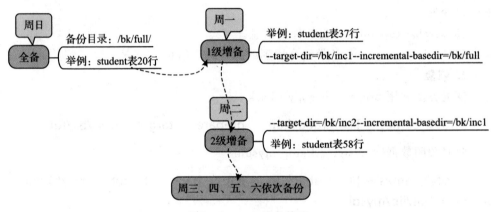

图 8-2 PXB 增备策略

1. 一级增量备份和恢复

（1）备份

[root@jeames ~]# xtrabackup -uroot -proot -S/var/lib/mysql/mysql.sock \
--backup --target-dir=/bk/inc1 --incremental-basedir=/bk/full --no-server-version-check

no-server-version-check 此参数控制是否将 MySQL 系统版本与 Percona XtraBackup 版本进行比较

（2）恢复

[root@jeames ~]# xtrabackup --prepare --apply-log-only --target-dir=/bk/full/

[root@jeames ~]# xtrabackup --prepare --apply-log-only \

-- target - dir= /bk/full/ -- incremental - dir= /bk/inc1

--此时关闭数据库 systemctl stop mysqld

[root@jeames ~]# xtrabackup -- copy - back -- target - dir= /bk/full -- datadir= /var/lib/mysql

[root@jeames ~]# chown - R mysql:mysql /var/lib/mysql

2. 二级增量备份和恢复

（1）备份

[root @ jeames ~] # xtrabackup - uroot - proot - S/var/lib/mysql/mysql. sock \ -- backup -- target - dir= /bk/inc2 \

-- incremental - basedir= /bk/inc1 -- no - server - version - check

（2）恢复

--此时关闭数据库 systemctl stop mysqld

[root@jeames ~]# xtrabackup -- prepare \

-- apply - log - only -- target - dir= /bk/full/ \

-- incremental - dir= /bk/inc2

--删除数据目录,恢复

[root@jeames ~]# xtrabackup -- copy - back \

-- target - dir= /bk/full/ -- datadir= /var/lib/mysql

[root@jeames ~]# chown - R mysql:mysql /var/lib/mysql

第 9 章 MySQL 误操作恢复

9.1 Mysqlbinlog 恢复误删除

9.1.1 建表

mysql> create database jeames

mysql> use jeames

CREATE TABLE 'tb1' (
'id' int(10) NOT NULL AUTO_INCREMENT
'name' char(10) CHARACTER SET latin1 DEFAULT NULL
PRIMARY KEY ('id')
) ENGINE = InnoDB DEFAULT CHARSET = utf8

mysql> insert into tb1 (name) value ('aa'),('bb')
mysql> show master status

File	Position	Binlog_Do_DB	Binlog_Ignore_DB	Executed_Gtid_Set
binlog.000028	978			

9.1.2 全备

[root@jeames ~]# mkdir /dakup
[root@jeames ~]# mysqldump -uroot -proot --hex-blob --routines \
--events --triggers --master-data=2 --single-transaction \
--databases jeames > /dakup/jeames.sql
##插入数据
mysql> insert into tb1 (name) value ('cc'),('dd')
mysql> flush logs
mysql> insert into tb1 (name) value ('ee')

```
mysql> show master status
+--------------+----------+--------------+------------------+-------------------+
| File         | Position | Binlog_Do_DB | Binlog_Ignore_DB | Executed_Gtid_Set |
+--------------+----------+--------------+------------------+-------------------+
| binlog.000029|      443 |              |                  |                   |
+--------------+----------+--------------+------------------+-------------------+

mysql> select * from tb1
+----+------+
| id | name |
+----+------+
|  1 | aa   |
|  2 | bb   |
|  3 | cc   |
|  4 | dd   |
|  5 | ee   |
+----+------+
```

9.1.3 误删除,删除数据库

```
mysql> drop database jeames
```

9.1.4 全备恢复

```
mysql> source /dakup/jeames.sql
mysql> select * from jeames.tb1
+----+------+
| id | name |
+----+------+
|  1 | aa   |
|  2 | bb   |
+----+------+
```

此时发现只恢复到了全部的数据记录。

```
--切换到binlog目录执行
mysqlbinlog -v --base64-output=DECODE-ROWS binlog.000028 \
| grep -C 6 -i "drop database"
mysqlbinlog -v --base64-output=DECODE-ROWS binlog.000029 \
| grep -C 6 -i "drop database"
```

——确认事务结束的位置，确认为 443

mysqlbinlog -- start-position=978 -- stop-position=630 \
binlog.000028 binlog.000029 -vv --database=jeames --skip-gtids | more

binlog 日志如图 9-1 所示。

```
# at 312
#221004 21:54:20 server id 1  end_log_pos 369 CRC32 0x4b496180  Table_map: `jeames`.`tb1` mapped to number 103
# at 369
#221004 21:54:20 server id 1  end_log_pos 412 CRC32 0x89727152  Write_rows: table id 103 flags: STMT_END_F
BINLOG '
DDs8YxMBAAAAOQAAAHEBAAAAGCAAAAAAAEABmplYW1lcwADdGIxAAID/gL+CgIBAQACAQiAYU1L
DDs8Yx4BAAAAKWAAAJwBAAAAGCAAAAAAAEAAgAC/wAFAAAAAmV1UnFyiQ==
/*!*/;
### INSERT INTO `jeames`.`tb1`
### SET
###   @1=5 /* INT meta=0 nullable=0 is_null=0 */
###   @2='ee' /* STRING(10) meta=65034 nullable=1 is_null=0 */
# at 412
#221004 21:54:20 server id 1  end_log_pos 443 CRC32 0x57444f24  Xid = 67
COMMIT/*!*/;
# at 443
# at 520
#221004 21:54:50 server id 1  end_log_pos 630 CRC32 0xe18ce670  Query   thread_id=9   exec_time=0   error_code=0    X
d = 70
SET TIMESTAMP=1664891690/*!*/;
drop database jeames
/*!*/;
DELIMITER ;
# End of log file
/*!50003 SET COMPLETION_TYPE=@OLD_COMPLETION_TYPE*/;
/*!50530 SET @@SESSION.PSEUDO_SLAVE_MODE=0*/;
[root@jeames mysql]#
```

图 9-1 binlog 日志

从图 9-1 的 binlog 记录可以看到增量变化：
开始位置从文件 binlog.000028,978
结束位置到文件 binlog.000029,430

9.1.5 binlog 恢复

mysqlbinlog -- start-position=978 -- stop-position=430 \
binlog.000028 binlog.000029 --database=jeames \
--skip-gtids -D | mysql -uroot -proot jeames

9.1.6 检验数据

mysql> select * from jeames.tb1;

```
+----+------+
| id | name |
+----+------+
|  1 | aa   |
|  2 | bb   |
|  3 | cc   |
|  4 | dd   |
|  5 | ee   |
+----+------+
```

此时发现数据已经恢复。

9.2 第三方工具

9.2.1 binlog2sql

工具官网:https://github.com/danfengcao/binlog2sql

从 MySQL binlog 解析出我们要的 SQL。根据不同选项,可以得到原始 SQL、回滚 SQL、去除主键的 INSERT SQL 等,主要用途如下:

(1)数据快速回滚(闪回)。
(2)主从切换后新 master 丢数据的修复。
(3)从 binlog 生成标准 SQL,带来的衍生功能。

9.2.2 my2sql

工具官网:https://github.com/liuhr/my2sql。

go 版 MySQL binlog 解析工具,通过解析 MySQL binlog,可以生成原始 SQL、回滚 SQL、去除主键的 INSERT SQL 等,也可以生成 DML 统计信息。类似工具有 binlog2sql、MyFlash、my2fback 等,本工具基于 my2fback、binlog_rollback 工具二次开发而来,主要用途如下:

(1)数据快速回滚(闪回)。
(2)主从切换后新 master 丢数据的修复。
(3)从 binlog 生成标准 SQL,带来的衍生功能。
(4)生成 DML 统计信息,可以找到哪些表更新得比较频繁。
(5)IO 高 TPS 高,查出哪些表在频繁更新。
(6)找出某个时间点数据库是否有大事务或者长事务。
(7)主从延迟,分析主库执行的 SQL 语句。
(8)除了支持常规数据类型,对大部分工具不支持的数据类型做了支持,比如 json、blob、text、emoji 等数据类型 sql 生成。

9.2.3 MyFlash

工具官网:https://github.com/Meituan-Dianping/MyFlash。

MyFlash 是由美团点评公司技术工程部开发维护的一个回滚 DML 操作的工具。该工具通过解析 v4 版本的 binlog,完成回滚操作。相对已有的回滚工具,其增加了更多的过滤选项,让回滚更加容易。该工具已经在美团点评内部使用。

Part 5　高可用

第 10 章 主从复制

10.1 主从复制简介

MySQL 主从复制是一个异步的复制过程,底层是基于 Mysql 数据库自带的二进制日志功能。就是一台或多台 MySQL 数据库(slave,即从库)从另一台 MySQL 数据库(master,即主库)进行日志的复制,然后再解析日志并应用到自身,最终实现从库的数据和主库的数据保持一致。MySQL 主从复制是 MySQL 数据库自带功能,无须借助第三方工具。

10.2 主从复制优点

MySQL 主从复制的优点主要包括以下 3 个方面:
(1)如果主库出现问题,可以快速切换到从库提供服务。
(2)可以在从库上执行查询操作,降低主库的访问压力。
(3)可以在从库上执行备份,以避免备份期间影响主库的服务。
MySQL 的 Replication 是一个多 MySQL 数据库做主从同步的方案,特点是异步复制,广泛用在各种对 MySQL 有更高性能、更高可靠性要求的场合,主从复制有以下几方面的好处:
(1)数据备份(Data Backup)。只是简单对数据库进行备份,降低数据丢失的风险。
(2)线下统计。用于报表等对数据时效性要求不高的场合。
(3)负载均衡(Load Balance)、读写分离。主要用在 MySQL 集群,解决单点故障或做故障切换,以降低单台服务器的负载和风险,如实现读写分离,可以使得服务器访问负荷比较均衡。
(4)数据分发(Data DistributIOn)、灾备。主要用于多数据中心或异地备份,实现数据分发与同步。
(5)高可用和数据容错(High Availability and Failover)。MySQL 自带的健康监控和检测,根据配置的时间间隔,可以检测主库是否正常工作,一旦发现主库宕机或无法正常工作,就会选择到最好的一个备库上。

10.3 主从复制原理

MySQL 主从复制原理如图 10-1 所示。

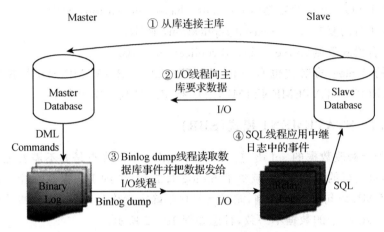

图 10-1　MySQL 主从复制原理

要实施复制,首先必须打开 Master 端的 binary log(bin-log)功能,否则无法实现。因为整个复制过程实际上就是 Slave 从 Master 端获取该日志然后再在自己身上完全顺序的执行日志中所记录的各种操作。

MySQL 复制的基本过程如下:

(1)Slave 上面的 IO 线程连接上 Master,并请求从指定日志文件的指定位置(或者从最开始的日志)之后的日志内容。

(2)Master 接收到来自 Slave 的 IO 线程的请求后,通过负责复制的 IO 线程根据请求信息读取指定日志指定位置之后的日志信息,返回给 Slave 端的 IO 线程。返回信息中除了日志所包含的信息之外,还包括本次返回的信息在 Master 端的 Binary Log 文件的名称以及在 Binary Log 中的位置。

(3)Slave 的 IO 线程接收到信息后,将接收到的日志内容依次写入到 Slave 端的 Relay Log 文件(mysql-relay-bin.xxxxxx)的最末端,并将读取到的 Master 端的 binlog 的文件名和位置记录到 master-info 文件中,以便在下一次读取的时候能够清楚地告诉 Master:我需要从某个 bin-log 的哪个位置开始往后的日志内容,请发给我。

(4)Slave 的 SQL 线程检测到 Relay Log 中新增加了内容后,会马上解析该 Log 文件中的内容成为在 Master 端真实执行时候的那些可执行的 SQL 语句,并在自身执行这些 SQL。这样,实际上就是在 Master 端和 Slave 端执行了同样的 SQL,所以两端的数据是完全一样的。

10.4 主从复制方式

MySQL 复制主要有三种方式，建议使用 MBR：

(1)基于 SQL 语句的复制(statement - based replication，SBR)。

(2)基于行的复制(row - based replication，RBR)。

(3)混合模式复制(mixed - based replication，MBR)。

对应的，binlog 的格式也有三种：STATEMENT，ROW，MIXED，由参数 binlog_format={ROW|STATEMENT|MIXED} 指定二进制日志的类型。

10.4.1 STATEMENT 模式(SBR)

每一条会修改数据的 sql 语句会记录到 binlog 中。优点是并不需要记录每一条 sql 语句和每一行的数据变化，减少了 binlog 日志量，节约 IO，提高性能。缺点是不是所有的 DML 语句都能被复制，尤其是包含不确定操作的时候，在某些情况下会导致 master - slave 中的数据不一致，日志如图 10 - 2 所示。

```
#220417 10:11:41 server id 80273306  end_log_pos 235 CRC32 0x78a96ecc   Anonymous_GTID  last_committed=0
=1      rbr_only=no     original_committed_timestamp=1650161501247690   immediate_commit_timestamp=165016
tion_length=330
# original_commit_timestamp=1650161501247690 (2022-04-17 10:11:41.247690 CST)
# immediate_commit_timestamp=1650161501247690 (2022-04-17 10:11:41.247690 CST)
/*!80001 SET @@session.original_commit_timestamp=1650161501247690*//*!*/;
/*!80014 SET @@session.original_server_version=80027*//*!*/;
/*!80014 SET @@session.immediate_server_version=80027*//*!*/;
SET @@SESSION.GTID_NEXT= 'ANONYMOUS'/*!*/;
# at 235
#220417 10:11:41 server id 80273306  end_log_pos 330 CRC32 0x52bf9853   Query   thread_id=7     exec_time
SET TIMESTAMP=1650161501/*!*/;
SET @@session.pseudo_thread_id=7/*!*/;
SET @@session.foreign_key_checks=1, @@session.sql_auto_is_null=0, @@session.unique_checks=1, @@session.au
SET @@session.sql_mode=1168113696/*!*/;
SET @@session.auto_increment_increment=1, @@session.auto_increment_offset=1/*!*/;
/*!\C utf8mb4 *//*!*/;
SET @@session.character_set_client=255,@@session.collation_connection=255,@@session.collation_server=255/
SET @@session.lc_time_names=0/*!*/;
SET @@session.collation_database=DEFAULT/*!*/;
/*!80011 SET @@session.default_collation_for_utf8mb4=255*//*!*/;
BEGIN
/*!*/;
# at 330
#220417 10:11:41 server id 80273306  end_log_pos 455 CRC32 0xa9158ff3   Query   thread_id=7     exec_time
use ceshi2 /*!*/;
SET TIMESTAMP=1650161501/*!*/;
update articles set content='it007'
/*!*/;
# at 455
#220417 10:11:41 server id 80273306  end_log_pos 486 CRC32 0xf6ab7400   Xid = 23
COMMIT/*!*/;
SET @@SESSION.GTID_NEXT= 'AUTOMATIC' /* added by mysqlbinlog */ /*!*/;
DELIMITER ;
# End of log file
/*!50003 SET COMPLETION_TYPE=@OLD_COMPLETION_TYPE*/;
/*!50530 SET @@SESSION.PSEUDO_SLAVE_MODE=0*/;
[root@jeames mysql-bin]#
```

图 10 - 2 binlog 的 STATEMENT 模式

STATEMENT 模式的优点：

(1)历史悠久，技术成熟。

(2)binlog 文件较小。

(3) binlog 中包含了所有数据库更改信息,可以据此来审核数据库的安全等情况。

(4) binlog 可以用于实时的还原,而不仅用于复制。

(5) 主从版本可以不一样,从服务器版本可以比主服务器版本高。

STATEMENT 模式的缺点:

(1) 使用以下函数的语句无法被复制。

① LOAD_FILE()。

② UUID()。

③ USER()。

④ FOUND_ROWS()。

⑤ SYSDATE()(除非启动时启用了 -- sysdate - is - now 选项)。

(2) INSERT ... SELECT 会产生比行的复制模式更多的行级锁。

(3) 复制需要进行全表扫描(WHERE 语句中没有使用到索引)的 DML 语句时,需要比行的复制模式 请求更多的行级锁。

(4) 对于有 AUTO_INCREMENT 字段的 InnoDB 表而言,INSERT 语句会阻塞其他 INSERT 语句。

(5) 对于一些复杂的语句,在从服务器上的耗资源情况会更严重,而行的复制模式下,只会对那个发生变化的记录产生影响。

(6) 执行复杂语句如果出错的话,会消耗更多资源。

10.4.2 ROW 模式(RBR)

不记录每条 sql 语句的上下文信息,仅需记录哪条数据被修改了,修改成什么样了。而且不会出现某些特定情况下的存储过程、或 function、或 trigger 的调用和触发无法被正确复制的问题。缺点是会产生大量的日志,尤其是 alter table 的时候会让日志暴涨,日志模式如图 10 - 3 所示。

1. ROW 模式的优点:

(1) 多数情况下,从服务器上的表如果有主键的话,复制就会快了很多。

(2) 复制以下几种语句时的行锁更少。

① INSERT ... SELECT。

② 包含 AUTO_INCREMENT 字段的 INSERT。

③ 没有附带条件或者并没有修改很多记录的 UPDATE 或 DELETE 语句。

(3) 执行 INSERT,UPDATE,DELETE 语句时锁更少。

(4) 从服务器上采用多线程来执行复制成为可能。

(5) 可以将 InnoDB 的事务隔离基本设为 READ COMMITTED,以获得更好的并发性。

```
### UPDATE ceshi2 . articles
### WHERE
###   @1=11 /* INT meta=0 nullable=0 is_null=0 */
###   @2='hahahahahaha' /* LONGBLOB/LONGTEXT meta=4 nullable=0 is_null=0 */
### SET
###   @1=11 /* INT meta=0 nullable=0 is_null=0 */
###   @2='jeames' /* LONGBLOB/LONGTEXT meta=4 nullable=0 is_null=0 */
### UPDATE ceshi2 . articles
### WHERE
###   @1=12 /* INT meta=0 nullable=0 is_null=0 */
###   @2='xixixixixix' /* LONGBLOB/LONGTEXT meta=4 nullable=0 is_null=0 */
### SET
###   @1=12 /* INT meta=0 nullable=0 is_null=0 */
###   @2='jeames' /* LONGBLOB/LONGTEXT meta=4 nullable=0 is_null=0 */
### UPDATE ceshi2 . articles
### WHERE
###   @1=13 /* INT meta=0 nullable=0 is_null=0 */
###   @2='aiaiaiaia' /* LONGBLOB/LONGTEXT meta=4 nullable=0 is_null=0 */
### SET
###   @1=13 /* INT meta=0 nullable=0 is_null=0 */
###   @2='jeames' /* LONGBLOB/LONGTEXT meta=4 nullable=0 is_null=0 */
### UPDATE ceshi2 . articles
### WHERE
###   @1=14 /* INT meta=0 nullable=0 is_null=0 */
###   @2='hohoahaoaooo' /* LONGBLOB/LONGTEXT meta=4 nullable=0 is_null=0 */
### SET
###   @1=14 /* INT meta=0 nullable=0 is_null=0 */
###   @2='jeames' /* LONGBLOB/LONGTEXT meta=4 nullable=0 is_null=0 */
# at 560
#220417 10:08:15 server id 80273306  end_log_pos 591 CRC32 0x35abb90d    Xid = 17
COMMIT/*!*/;
SET @@SESSION.GTID_NEXT= 'AUTOMATIC' /* added by mysqlbinlog */ /*!*/;
DELIMITER ;
```

图 10-3 binlog 的 ROW 模式

2. ROW 模式的缺点：

(1)binlog（二进制日志）非常大。

(2)复杂的回滚时 binlog 中会包含大量的数据。

(3)主服务器上执行 DML 语句时，所有发生变化的记录都会写到 binlog 中，而 STATEMENT 模式只会写一次，这会导致频繁发生 binlog 的并发写问题。

(4)UDF 产生的大 BLOB 值会导致复制变慢。

(5)无法从 binlog 中看到都复制了写什么语句。

10.4.3　MIXED 模式(MBR)

STATEMENT 模式和 ROW 模式的混合使用，一般的复制使用 STATEMENT 模式保存 binlog，对于 STATEMENT 模式无法复制的操作使用 ROW 模式保存 binlog，MySQL 会根据执行的 SQL 语句选择日志保存方式。

binlog 模式修改：

(1)查询 binlog 模式

mysql> select @@binlog_format

(2)设定 binlog 模式

set session binlog_format = STATEMENT

(3)确认模式

mysql> show variables like '%log_bin%'

(4) 解析 binlog

mysqlbinlog mysql-bin.000006 -vvv

10.5 传统异步主从复制部署

10.5.1 主库参数配置

```
cat > /mysqljem/master/my.cnf << "EOF"
[mysqld]
user= mysql
port= 3306
character_set_server= utf8mb4
secure_file_priv= ''
server-id = 802733261
log-bin =
binlog_format= row
binlog-ignore-db = mysql
binlog-ignore-db = information_schema
binlog-ignore-db = performance_schema
binlog-ignore-db = sys
replicate_ignore_db= information_schema
replicate_ignore_db= performance_schema
replicate_ignore_db= mysql
replicate_ignore_db= sys
skip-name-resolve
report_host= 172.72.0.72
EOF
```

10.5.2 从库参数配置

```
cat > /mysqljem/slave/my.cnf << "EOF"
[mysqld]
user= mysql
port= 3306
character_set_server= utf8mb4
secure_file_priv= ''
```

```
server-id = 802733262
log-bin =
binlog_format= row
binlog-ignore-db = mysql
binlog-ignore-db = information_schema
binlog-ignore-db = performance_schema
binlog-ignore-db = sys
replicate_ignore_db= information_schema
replicate_ignore_db= performance_schema
replicate_ignore_db= mysql
replicate_ignore_db= sys
skip_name_resolve
report_host= 172.72.0.73
EOF
```

注意：

(1) binlog-ignore-db：此参数表示不记录指定的数据库的二进制日志。

(2) replicate_ignore_db：用来设置不需要同步的库。

10.5.3 主库创建用户

mysql> create user repl@'172.72.0.73' identified with mysql_native_password by 'root';

mysql> grant all on *.* torepl@'172.72.0.73' with grant option;

mysql> flush privileges;

注意：172.72.0.73 为从库 ip。

10.5.4 主从数据同步

【主库 dump 导出 jemdb 数据库】

```
mysqldump -uroot -proot -h172.72.0.72 -P3306 \
--single-transaction \
--hex-blob --routines --events --triggers --source-data=2 \
--set-gtid-purged=OFF --databases jemdb --default-character-set=utf8 \
--max_allowed_packet=512M > /tmp/salve.sql
```

注意：如果是所有数据库，就是-A。

以下在主库操作插入数据：

mysql> show master status \G

```
mysql>  insert into jemdb.mytb1 values(3,'a'),(4,'b')
mysql>  show slave hosts
mysql>  select @@hostname,@@server_id,@@server_uuid
```

【从库同步主库数据】

(1)通过以下方法查找同步的启示位置。

```
[root@jeames ~ ]# grep "CHANGE MASTER" /tmp/salve.sql
-- CHANGE MASTER TO MASTER_LOG_FILE= 'master-bin.000002', MASTER_LOG_POS= 1144
```

(2)dump 导入库

```
mysql -uroot -proot -h172.72.0.72 -P3306 < /tmp/salve.sql
```

(3)登录从库同步。

```
[root@jeames ~ ]# mysql -uroot -proot -h172.72.0.73 -P3306
mysql>  change master to
master_host= '172.72.0.72'
master_user= 'repl'
master_password= 'root'
master_port= 3306
master_log_file= 'master-bin.000002'
master_log_pos= 1144
```

注意：master-bin.000002 为主库 binlog，1144 为 binlog 中的 MASTER_LOG_POS。

(4)从库同步开始

```
mysql>  start slave;
```

【主库确认】

```
mysql>  show slave hosts
```

如果同步 OK，则可以看到图 10-4 所示的信息。

```
mysql> show slave hosts;
+-----------+-------------+------+-----------+--------------------------------------+
| Server_id | Host        | Port | Master_id | Slave_UUID                           |
+-----------+-------------+------+-----------+--------------------------------------+
| 802733262 | 172.72.0.73 | 3306 | 802733261 | 10b05a93-bd91-11ec-baec-0242ac480049 |
+-----------+-------------+------+-----------+--------------------------------------+
1 row in set, 1 warning (0.00 sec)
```

图 10-4 MySQL 主从同步模式

【从库设置只读】

mysql> show variables like '%read%';

mysql> set global read_only = 1;

10.6 主从复制维护

10.6.1 主从状态查询

--从库查询

mysql> show slave status \G

如图 10-5 所示,证明目前没有主从延迟状态。

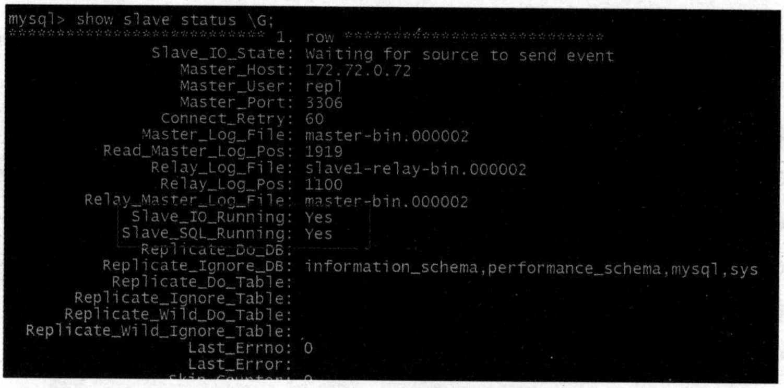

图 10-5 MySQL 主从同步从库状态

Master_Log_File = Relay_Master_Log_File

Read_Master_Log_Pos = Exec_Master_Log_Pos

--主库查询,主库状态(图 10-6)

mysql> show master status

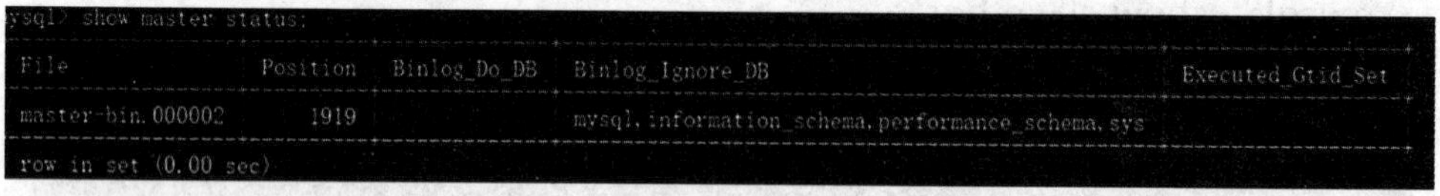

图 10-6 MySQL 主从同步主库状态

--参数说明

#从库的延迟

Seconds_Behind_Master: 0

```
#从库上 I/O thread 负责请求和接收主库传递来的 binlog 信息
Slave_IO_Running
#从库上 SQL thread 负责应用 relay 中的 binlog 的信息
Slave_SQL_Running
```

10.6.2 主从复制异常排查

1. 线程查询

```
SELECT *
FROM performance_schema.threads a
WHERE a.'NAME' IN ( 'thread/sql/slave_IO', 'thread/sql/slave_sql' ) or a.PROCESSLIST_COMMAND= 'Binlog Dump'
SELECT * FROM information_schema.'PROCESSLIST' a
where a.USER= 'system user' or a.command= 'Binlog Dump'
```

2. IO_thread 异常

IO_thread 异常,状态往往是 Slave_IO_Running:Connecting 或 NO

IO_thread 是向 Master 发送请求读取 master binlog,如果处于 Connecting 状态,说明无法正确地与 Master 进行连接

3. sql_thread 异常

sql_thread 发生异常,状态就会变为 Slave_SQL_Running:NO

sql_thread 发生异常的情况非常多,发生异常后,需要通过从库排查和解决:
对比主库和从库的二进制日志的情况
通过 show slave status\G 查看错误信息
根据这些报错信息,往往就能够定位到发生异常的原因

10.6.3 主键错误导致延迟及解决

```
--主库
mysql> use jemdb
mysql> create table test_tb(id int primary key ,name varchar(30))
mysql> insert into test_tb values(1,'a'),(2,'b')
--从库
mysql> insert into test_tb values(3,'c'),(4,'d')
--主库
mysql> insert into test_tb values(3,'c0'),(4,'d0')
mysql> flush logs
```

――从库

mysql> show slave status \G

此时我们发现 test_tb 表主键有重复了,则同步会报错

mysql> select * from performance_schema.replication_applier_status_by_worker

如果我们了解产生异常的具体事件,而且能够掌控,可以通过设置 sql_slave_skip_counter 参数来跳过当前错误。

――如果是 10,就是跳过接下来的 10 个错误

mysql> Set global sql_slave_skip_counter= 1

mysql> start slave sql_thread

或者使用 slave_skip_errors 参数(read only variable),指定跳过某种类型的错误,参数文件中设置:slave_skip_errors = 1062 #跳过 1062 错误。

10.7 单主 2 从 GTID 复制

GTID(Global Transaction ID,全局事务 ID)是全局事务标识符,是一个已提交事务的编号,并且是一个全局唯一的编号。

GTID 是从 MySQL 5.6 版本开始在主从复制方面推出的重量级特性。

GTID 实际上是由 UUID+TID 组成的。其中 UUID 是一个 MySQL 实例的唯一标识。

GTID 代表了该实例上已经提交的事务数量,并且随着事务提交单调递增。

GTID 有如下几点作用:

(1)根据 GTID 可以知道事务最初是在哪个实例上提交的。

(2)GTID 的存在方便了 Replication 的 Failover。因为不用像传统模式复制那样去找 master_log_file 和 master_log_pos。

(3)基于 GTID 搭建主从复制更加简单,确保每个事务只会被执行一次。

GTID 的工作原理为:

(1)slave 端的 i/o 线程将变更的 binlog,写入到本地的 relay log 中。

(2)sql 线程从 relay log 中获取 GTID,然后对比 slave 端的 binlog 是否有记录。

(3)如果有记录,说明该 GTID 的事务已经执行,slave 会忽略。

(4)如果没有记录,slave 就会从 relay log 中执行该 GTID 的事务,并记录到 binlog。

(5)在解析过程中会判断是否有主键,如果没有就用二级索引,如果没有就用全部扫描。

10.7.1 主库参数配置

cat /mysqlgtid/master/my.cnf<< "EOF"

```
[mysqld]
port=3306
character_set_server=utf8mb4
secure_file_priv=
server-id=802733061
log-bin=
binlog_format=row
binlog-ignore-db = mysql
binlog-ignore-db = information_schema
binlog-ignore-db = performance_schema
binlog-ignore-db = sys
replicate_ignore_db=information_schema
replicate_ignore_db=performance_schema
replicate_ignore_db=mysql
replicate_ignore_db=sys
log-slave-updates=1
skip-name-resolve
gtid-mode=ON
enforce-gtid-consistency=on
report_host=172.72.0.5
EOF
```

10.7.2 从库参数配置

```
cat> /mysqlgtid/slave1/my.cnf<< "EOF"
[mysqld]
port=3306
character_set_server=utf8mb4
secure_file_priv=
server-id=802733062
log-bin=
binlog_format=row
binlog-ignore-db = mysql
binlog-ignore-db = information_schema
binlog-ignore-db = performance_schema
binlog-ignore-db = sys
replicate_ignore_db=information_schema
```

```
replicate_ignore_db=performance_schema
replicate_ignore_db=mysql
replicate_ignore_db=sys
log-slave-updates=1
gtid-mode=ON
enforce-gtid-consistency=ON
skip_name_resolve
report_host=172.72.0.6
EOF

cat>/mysqlgtid/slave2/conf.d/my.cnf<<"EOF"
[mysqld]
user=mysql
port=3306
character_set_server=utf8mb4
secure_file_priv=
server-id=802733063
log-bin=
binlog_format=row
binlog-ignore-db=mysql
binlog-ignore-db=information_schema
binlog-ignore-db=performance_schema
binlog-ignore-db=sys
replicate_ignore_db=information_schema
replicate_ignore_db=performance_schema
replicate_ignore_db=mysql
replicate_ignore_db=sys
log-slave-updates=1
gtid-mode=ON
enforce-gtid-consistency=ON
skip_name_resolve
report_host=172.72.0.7
EOF
```

(1) GTID 不需要传统的 binlog 和 position 号了, 而是在从库"change master to"时使用"master_auto_position=1"的方式搭建。

(2) 在启用基于 GTID 的复制之前,必须将变量 enforce-gtid-consistency 设置

为 ON。

10.7.3 主库创建用户

mysql> create user repl@'%' identified with mysql_native_password by 'root'
mysql> grant all on *.* to repl@'%' with grant option
mysql> flush privileges
mysql> select user,host,grant_priv,password_last_changed,authentication_string from mysql.user

--创建数据库
mysql> create database jemdb
mysql> use jemdb
mysql> create table mytb1(id int,name varchar(30))
mysql> insert into mytb1 values(1,'a'),(2,'b')
mysql> select * from mytb1

10.7.4 dump 导出主库

[root@jeames ~]# mysqldump -uroot -proot -h172.72.0.5 -P3306 \
--single-transaction \
--hex-blob --routines --events --triggers --source-data=2 \
--databases jemdb --default-character-set=utf8 \
--max_allowed_packet=512M > /tmp/gtidsalve.sql

11.7.5 从库恢复应用

[root@jeames ~]# mysql -uroot -proot -h172.72.0.6 \
-P3306 < /tmp/gtidsalve.sql

mysql> change master to
master_host='172.72.0.5',
master_port=3306,
master_user='repl',
master_password='root',
master_auto_position=1

--启动应用
mysql> start slave

mysql> show slave hosts

以上为从库1的恢复应用,从库2设置同以上

10.7.6 从库设置只读

mysql> show variables like '%read%'
mysql> set global read_only= 1

第 11 章　MySQL Router 读写分离及负载均衡

11.1　MySQL Router 简介

MySQL Router 的主要用途是读写分离，主主故障自动切换，负载均衡，连接池等。对于 MySQL 强烈建议使用 Router 8 与 MySQL Server8 和 5.7 一起使用，MySQL Router 是 MySQL 官方提供的一个轻量级中间件，可以在应用程序与 MySQL 服务器之间提供透明的路由方式。主要用以解决 MySQL 主从库集群的高可用、负载均衡、易扩展等问题。

11.1.1　优点

(1) 实现了主从读写分离。
(2) 读从库一定程度上起到负载均衡的作用。
(3) 可以当作连接池。
(4) 主主架构中，主数据库发生故障后，主从 MySQL 服务器自动发生切换。
(5) 多从服务器负载读时，其中一台从服务器 DOWN，不影响业务访问。

11.1.2　缺点

(1) 如果从库延时或因故障同步停止，那么对提交事务即读的业务来并不适用。
(2) MySQL Router 是中间代理，有一定的性能损耗。
(3) 运维管理并不完善，对于主主切换，主从负载均衡管理并不方便。

11.2　MySQL Router 工作流程

MySQL Router 的工作流程如下：
(1) MySQL 客户端或连接器连接到 MySQL Router，例如端口 7001。
(2) MySQL Router 检查可用的 MySQL 服务器。
(3) MySQL Router 开启一个适当的 MySQL 服务器的连接。
(4) MySQL Router 在应用程序和 MySQL 服务器之间来回转发数据包。
(5) 如果连接的 MySQL 服务器发送故障，MySQL Router 将断开应用程序的连接。然后，应用程序可以重新尝试连接到 MySQL Router，MySQL Router，选择一个新的可用的 MySQL 服务器。

11.3 读写分离 + 负载均衡

以下为基于 Centos8 系统上运行 MySQL8 的场景,读写分离 + 负载均衡的架构如图 11-1 所示。

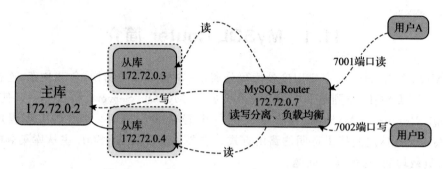

图 11-1 MySQL Router 读写分离 + 负载均衡

11.3.1 搭建 1 主 2 从

大家可按照以下表 11-1 的规划进行部署。

表 11-1 MySQL Router 读写分离 + 负载均衡服务器规划

角色	ip 地址	主机名	server_id	类型
master	172.72.0.2	master	802733062	主库
slave	172.72.0.3	slave1	802733063	从库 1
slave	172.72.0.4	Slave2	802733064	从库 2
中间件	172.72.0.7	MysqlRouter	—	读写分离、负载均衡

1. 主库参数

cat > /etc/my.cnf <<"EOF"
[mysqld]
port = 3306
character_set_server = utf8mb4
secure_file_priv =
server - id = 802733062
log - bin =
binlog_format = row
binlog - ignore - db = mysql

```
binlog-ignore-db = information_schema
binlog-ignore-db = performance_schema
binlog-ignore-db = sys
skip-name-resolve
gtid-mode = ON
enforce-gtid-consistency = on
report_host = 172.72.0.2
EOF
```

2. 从库参数

```
cat > /etc/my.cnf <<"EOF"
[mysqld]
port = 3306
character_set_server = utf8mb4
secure_file_priv =
server-id = 802733063
log-bin =
binlog_format = row
replicate_ignore_db = information_schema
replicate_ignore_db = performance_schema
replicate_ignore_db = mysql
replicate_ignore_db = sys
gtid-mode = ON
enforce-gtid-consistency = ON
skip_name_resolve
report_host = 172.72.0.3
EOF

cat > /etc/my.cnf <<"EOF"
[mysqld]
user = mysql
port = 3306
character_set_server = utf8mb4
secure_file_priv =
server-id = 802733064
log-bin =
binlog_format = row
```

```
replicate_ignore_db = information_schema
replicate_ignore_db = performance_schema
replicate_ignore_db = mysql
replicate_ignore_db = sys
gtid-mode = ON
enforce-gtid-consistency = ON
skip_name_resolve
report_host = 172.72.0.4
EOF
```

11.3.2 主库创建复制用户

```
[root@jeames ~]# mysql -uroot -proot -h172.72.0.2
mysql> create database jeamesdb
mysql> use jeamesdb
mysql> create table mytb1(id int,name varchar(30))
mysql> insert into mytb1 values(1,'a'),(2,'b')
mysql> show variables like '%log_bin%'
mysql> create user repl@'%' identified with mysql_native_password by 'root'
mysql> grant all on *.* to repl@'%' with grant option
mysql> flush privileges
mysql> select @@hostname,@@server_id,@@server_uuid
```

11.3.3 从库应用

1. 登录 2 个从库

```
[root@jeames ~]# mysql -uroot -proot -h172.72.0.3 -P3306
[root@jeames ~]# mysql -uroot -proot -h172.72.0.3 -P3306
change master to
master_host = '172.72.0.2'
master_port = 3306
master_user = 'repl'
master_password = 'root'
master_auto_position = 1

mysql> start slave
```

mysql> show slave status \G

注意:2个从库按照上面做配置。

2. 登陆主库确认从库状态(图 11 - 2)

[root@jeames ~]# mysql - uroot - proot - h172.72.0.2 - P3306

mysql> show slave hosts

```
mysql> show slave hosts;
+------------+------------+------+------------+--------------------------------------+
| Server_id  | Host       | Port | Master_id  | Slave_UUID                           |
+------------+------------+------+------------+--------------------------------------+
| 802733064  | 172.72.0.4 | 3306 | 802733062  | 242376a4-c916-11ec-80f6-0242ac480004 |
| 802733063  | 172.72.0.3 | 3306 | 802733062  | defebc02-c915-11ec-9b0e-0242ac480003 |
+------------+------------+------+------------+--------------------------------------+
2 rows in set, 1 warning (0.00 sec)
```

图 11 - 2 MySQL 主从同步状态

3. 确认备库状态

[root@jeames ~]# mysql - uroot - proot - h172.72.0.4 - P3306

mysql> show slave statuS \G

11.3.4 MySQL Router 配置

MySQL Router 安装包到 MySQL 官方下载对应的版本即可。

1. 安装 MySQL Router

rpm - ivh mysql - router - community - 8.0.29 - 1.el8.x86_64.rpm

查看版本

mysqlrouter -- version

```
[root@MysqlRouter /]# rpm -ivh mysql-router-community-8.0.29-1.el8.x86_64.rpm
warning: mysql-router-community-8.0.29-1.el8.x86_64.rpm: Header V4 RSA/SHA256 Signature, key ID 3a79bd29:
Verifying...                          ################################# [100%]
Preparing...                          ################################# [100%]
Updating / installing...
   1:mysql-router-community-8.0.29-1.e################################# [100%]
```

图 11 - 3 安装 MySQL Router 配置

2. 目录创建

[root@MysqlRouter /]# mkdir - p /var/log/mysql - router

[root@MysqlRouter /]# mkdir - p /etc/mysql - router/

查看是否有 mysql 用户

[root@MysqlRouter /]# cat /etc/passwd

[root@MysqlRouter /]# useradd mysql

[root@MysqlRouter /]# chown mysql:mysql /var/log/mysql - router

3. 参数配置

```
cat > /etc/mysql-router/mysqlrouter.conf <<"EOF"
[DEFAULT]
logging_folder = /var/log/mysql-route
[logger]
level = INFO
#配置读操作
[routing:secondary]
bind_address = 172.72.0.7
#读操作的端口
bind_port = 7001
destinations = 172.72.0.3:3306,172.72.0.4:3306
routing_strategy = round-robin
#配置写操作
[routing:primary]
bind_address = 172.72.0.7
#写操作的端口
bind_port = 7002
destinations = 172.72.0.2:3306
routing_strategy = next-available
EOF
```

注意：round-robin 代表一种轮训算法

4. Router 启动

```
[root@MysqlRouter /]# mysqlrouter -c /etc/mysql-router/mysqlrouter.conf &
[root@MysqlRouter /]# netstat -anp | grep mysqlrouter
[root@MysqlRouter /]# ps -ef | grep mysqlrouter
```

5. 测试读写分离

(1) 测试读负载均衡 (7001 端口) (图 11-4)

```
[root@jeames ~]# for i in $(seq 1 10); do mysql -uroot -proot -h172.72.0.7 -P7001 -e 'select @@server_id;'; done | egrep '[0-9]'
```

(2) 测试写 (7002 端口) (图 11-5)

```
[root@jeames ~]# for i in $(seq 1 10); do mysql -uroot -proot -h172.72.0.7 -P7002 -e 'select @@server_id;'; done | egrep '[0-9]'
```

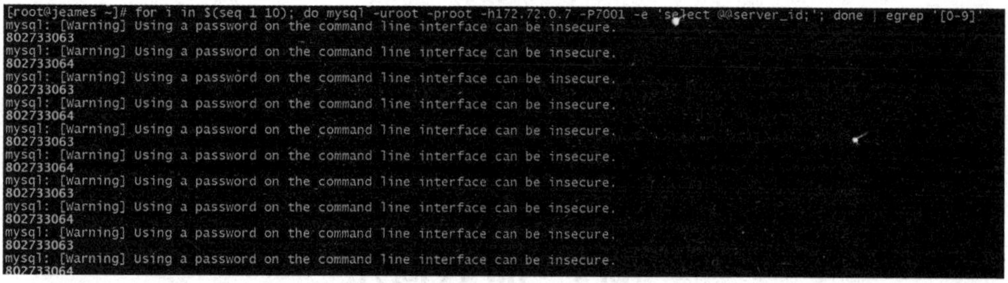

第 11 章　MySQL Router 读写分离及负载均衡

图 11 - 4　MySQL Router 读测试

图 11 - 5　MySQL Router 写测试

6. 从库设置为只读

mysql> show variables like '% read_only %'

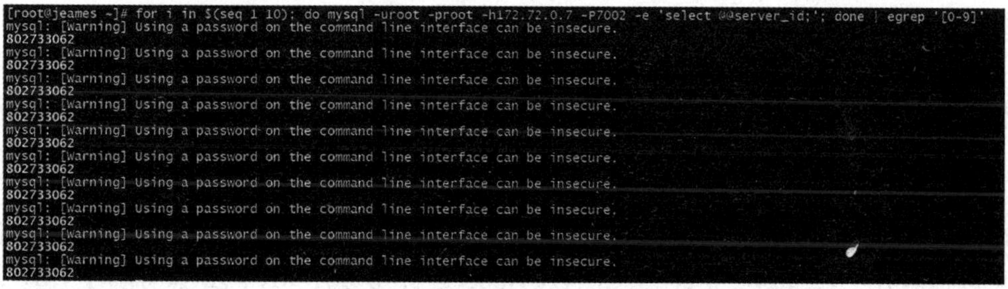

mysql> set global read_only = 1

mysql> set global super_read_only = 1

mysql> flush privileges

第 12 章 高可用 MHA 架构

12.1 MHA 简介

MHA（Master High Availability Manager and tools for MySQL）目前在 MySQL 高可用方面是一个相对成熟的解决方案，它是由 youshimaton 采用 Perl 语言编写的一个脚本管理工具。目前 MHA 主要支持一主多从的架构，搭建 MHA 要求一个复制集群最少有 3 台数据库服务器，一主二从，即一台充当 Master，一台充当备用 Master，另一台充当从库。

MHA 由两部分组成：MHA Manager（管理节点）和 MHA Node（数据库节点），MHA Manager 可以单独部署在一台独立的机器上管理多个 master-slave 集群，也可以部署在一台 slave 节点上。MHA Node 运行在每台 MySQL 服务器上，MHA Manager 会定时探测集群中的 master 节点，当 master 出现故障时，它可以自动将最新数据的 slave 提升为新的 master，然后将所有其他的 slave 重新指向新的 master。整个故障转移过程对应用程序完全透明，架构图如图 12-1 所示。

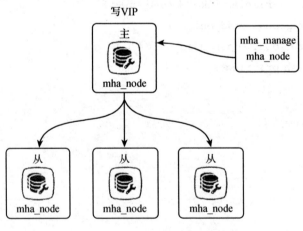

图 12-1 MHA 高可用架构图

12.2 架构规划

架构规划内容详见表 12-1 所列。

第 12 章 高可用 MHA 架构

表 12-1 架构规划内容

角色	ip 地址	主机名	server_id	类型
Monitor host	192.168.1.55	MHA-Monitor	-	监控复制组
Master	192.168.1.56	MHA-Master	1	写入(主)
Candicate master	192.168.1.57	MHA-Slave1	2	从库(主库的备用)
Slave	192.168.1.58	MHA-Slave2	3	从库

VIP:绑定到主库 192.168.1.54,主要目的是切服务器。

操作系统为:Centos7.3。

12.3 安装 MySQL8

以下 node 3 个节点需同时操作。

12.3.1 用户及组

groupadd mysql

useradd -r -g mysql mysql

12.3.2 解压缩安装包

tar -xf mysql-8.0.19-linux-glibc2.12-x86_64.tar.xz -C /usr/local/

ln -s /usr/local/mysql-8.0.19-linux-glibc2.12-x86_64 /usr/local/mysql8019

ln -s /usr/local/mysql8019 /usr/local/mysql

echo "export PATH=$PATH:/usr/local/mysql/bin" >> /etc/bashrc

source /etc/bashrc

chown -R mysql.mysql /usr/local/mysql-8.0.19-linux-glibc2.12-x86_64

12.3.3 yum 安装依赖

yum install -y net-tools

yum install -y libtinfo*

yum -y install numactl

yum -y install libaio*

yum -y install perl perl-devel

yum -y install autoconf

12.3.4 mysql 初始化

以下命令同一行执行。

/usr/local/mysql/bin/mysqld -- initialize - insecure -- user = mysql -- basedir = /usr/local/mysql -- datadir = /usr/local/mysql/data

12.3.5 mysql 自启动配置

cp /usr/local/mysql/support - files/mysql.server /etc/init.d/mysqld
chmod 755 /etc/init.d/mysqld
chkconfig -- add mysqld
chkconfig mysqld on
chkconfig -- level 345 mysqld on
systemctl restart mysqld
systemctl status mysqld

12.4 一主2从GTID同步

12.4.1 配置参数文件

1. Master1

cat > /etc/my.cnf <<"EOF"
[mysqld]
basedir = /usr/local/mysql
datadir = /usr/local/mysql/data
user = mysql
port = 3306
character_set_server = utf8mb4
secure_file_priv =
server - id = 803306131
log - bin =
binlog_format = row
binlog - ignore - db = mysql
binlog - ignore - db = information_schema
binlog - ignore - db = performance_schema
binlog - ignore - db = sys
replicate_ignore_db = information_schema
replicate_ignore_db = performance_schema
replicate_ignore_db = mysql
replicate_ignore_db = sys

```
log-slave-updates = 1
skip-name-resolve
log_timestamps = SYSTEM
#default-time-zone = '+8:00'
auto-increment-increment = 1
auto-increment-offset = 1
gtid-mode = ON
enforce-gtid-consistency = on
report_host = 192.168.1.56
EOF
```

2. Slave1

```
cat > /etc/my.cnf <<"EOF"
#S1
[mysqld]
basedir = /usr/local/mysql
datadir = /usr/local/mysql/data
user = mysql
port = 3306
character_set_server = utf8mb4
secure_file_priv =
server-id = 803306132
log-bin =
binlog_format = row
binlog-ignore-db = mysql
binlog-ignore-db = information_schema
binlog-ignore-db = performance_schema
binlog-ignore-db = sys
replicate_ignore_db = information_schema
replicate_ignore_db = performance_schema
replicate_ignore_db = mysql
replicate_ignore_db = sys
skip-name-resolve
log_timestamps = SYSTEM
#default-time-zone = '+8:00'
gtid-mode = ON
```

```
enforce-gtid-consistency=ON
report_host=192.168.1.57
EOF
```

3. Slave2

```
cat > /etc/my.cnf <<"EOF"
#S2
[mysqld]
basedir=/usr/local/mysql
datadir=/usr/local/mysql/data
user=mysql
port=3306
character_set_server=utf8mb4
secure_file_priv=
server-id = 803306133
log-bin =
binlog_format=row
binlog-ignore-db = mysql
binlog-ignore-db = information_schema
binlog-ignore-db = performance_schema
binlog-ignore-db = sys
replicate_ignore_db=information_schema
replicate_ignore_db=performance_schema
replicate_ignore_db=mysql
replicate_ignore_db=sys
skip-name-resolve
log_timestamps = SYSTEM
#default-time-zone='+8:00'
gtid-mode=ON
enforce-gtid-consistency=ON
report_host=192.168.1.58
EOF
```

12.4.2 主从同步

1. 主库操作

mysql> create user repl@'%' identified with mysql_native_password by 'root';

mysql> grant replication slave on *.* to repl@'%' with grant option;
mysql> create user mha@'%' identified with mysql_native_password by 'root';
mysql> grant all on *.* to 'mha'@'%' with grant option;
mysql> flush privileges;

mysql> show master status \G;
mysql> show slave hosts;
mysql> select @@server_id,@@server_uuid;

2. 从库操作

change master to
master_host='192.168.1.56',master_port=3306,master_user='repl',
master_password='root',master_auto_position=1;

mysql> start slave;
mysql> show slave status \G;

12.4.3　Master 绑定 VIP

#在主库上执行添加 VIP 的过程(第一次手动添加,后续启动切换)
[root@MHA-Master ~]# ifconfig
[root@MHA-Master ~]# /sbin/ifconfig ens33:1 192.168.1.54

12.5　互信设置

4 台机器互相免密码登录,应注意,自己跟自己也要配免密码登录。

12.5.1　在 Manager 上配置到所有的 Node 节点的无密码验证

[root@MHA-Monitor /]# ssh-keygen -t rsa
[root@MHA-Monitor /]# ssh-copy-id -i /root/.ssh/id_rsa.pub 'root@192.168.1.56'
[root@MHA-Monitor /]# ssh-copy-id -i /root/.ssh/id_rsa.pub 'root@192.168.1.57'
[root@MHA-Monitor /]# ssh-copy-id -i /root/.ssh/id_rsa.pub 'root@192.168.1.58'
[root@MHA-Monitor /]# ssh-copy-id -i /root/.ssh/id_rsa.pub 'root@192.168.1.55'

12.5.2 在 Master 上配置所有的 Node 节点的无密码验证

[root@MHA-Master ~]# ssh-keygen -t rsa

[root@MHA-Master ~]# ssh-copy-id -i /root/.ssh/id_rsa.pub 'root@192.168.1.55'

[root@MHA-Master ~]# ssh-copy-id -i /root/.ssh/id_rsa.pub 'root@192.168.1.56'

[root@MHA-Master ~]# ssh-copy-id -i /root/.ssh/id_rsa.pub 'root@192.168.1.57'

[root@MHA-Master ~]# ssh-copy-id -i /root/.ssh/id_rsa.pub 'root@192.168.1.58'

12.5.3 在 Candicate Master 上配置所有 Node 节点的无密码验证

[root@MHA-Slave1 /]# ssh-keygen -t rsa

[root@MHA-Slave1 /]# ssh-copy-id -i /root/.ssh/id_rsa.pub 'root@192.168.1.55'

[root@MHA-Slave1 /]# ssh-copy-id -i /root/.ssh/id_rsa.pub 'root@192.168.1.56'

[root@MHA-Slave1 /]# ssh-copy-id -i /root/.ssh/id_rsa.pub 'root@192.168.1.57'

[root@MHA-Slave1 /]# ssh-copy-id -i /root/.ssh/id_rsa.pub 'root@192.168.1.58'

12.5.4 在 Slave2 上配置所有 Node 节点的无密码验证

[root@MHA-Slave2 ~]# ssh-keygen -t rsa

[root@MHA-Slave2 ~]# ssh-copy-id -i /root/.ssh/id_rsa.pub 'root@192.168.1.55'

[root@MHA-Slave2 ~]# ssh-copy-id -i /root/.ssh/id_rsa.pub 'root@192.168.1.56'

[root@MHA-Slave2 ~]# ssh-copy-id -i /root/.ssh/id_rsa.pub 'root@192.168.1.57'

[root@MHA-Slave2 ~]# ssh-copy-id -i /root/.ssh/id_rsa.pub 'root@192.168.1.58'

12.5.5 测试 SSH

ssh 192.168.1.55 date

ssh 192.168.1.56 date
ssh 192.168.1.57 date
ssh 192.168.1.58 date

12.6 安装 MHA 软件

下载：https://github.com/yoshinorim/mha4mysql-manager/releases。

先安装 Node，再安装 Manager。

12.6.1 安装 MHA Node

在所有的 MySQL 服务器和 MHA Manager 主机上都安装 MHA Node。

yum install -y perl-DBD-MySQL perl-Config-Tiny perl-Log-Dispatch perl-Parallel-ForkManager perl-ExtUtils-CBuilder perl-ExtUtils-MakeMaker perl-CPAN

tar xf mha4mysql-node-0.58.tar.gz

cd mha4mysql-node-0.58

perl Makefile.PL

yum install make -y

make && make install

12.6.2 装 MHA Manager

MHA Manager 中主要几个管理员的命令行工具，也是依赖一些 Perl 模块的，只在管理节点安装即可。

tar zxf mha4mysql-manager-0.58.tar.gz

cd mha4mysql-manager-0.58

perl Makefile.PL   --先 NO，然后再 YES

make && make install

12.7 配置 MHA

（1）由于脚本中并没有 master_ip_failover 脚本，启动时会报错，也可以到 mha4mysqlmanager-0.5X/samples/scripts 下复制对应脚本到指定位置或注释掉 master_ip_failover_script。

（2）MHA 可以监控多个主从的集群，每个集群的配置文件可以用名字区分，因为这里只有一个集群，因此只有 mha.cnf 一个文件！

12.7.1　host 配置及创建目录

[root@MHA-Monitor ~]# more /etc/hosts
192.168.1.55 MHA-Monitor
192.168.1.56 MHA-Master
192.168.1.57 MHA-Slave1
192.168.1.58 MHA-Slave2

[root@MHA-Monitor /]# mkdir -p /usr/local/mha
[root@MHA-Monitor /]# mkdir -p /etc/mha

12.7.2　Manager 上创建配置文件

cat > /etc/mha/mha.cnf <<"EOF"
[server default]

1. 工作目录

manager_workdir = /usr/local/mha

2. 日志

manager_log = /usr/local/mha/manager_running.log

3. 自动切换脚本

master_ip_failover_script = /usr/local/mha/scripts/master_ip_failover

4. 手动切换脚本

master_ip_online_change_script = /usr/local/mha/scripts/master_ip_online_change

5. 检测频率

ping_interval = 1

6. 主从库配置，使用 host 及 IP

secondary_check_script = /usr/local/bin/masterha_secondary_check -s MHA-Slave1 -s MHA-Slave2 --user=root --master_host=MHA-Master --master_ip=192.168.1.55 --master_port=3306

7. mha 用户

ssh_user = root
user = mha

```
password = root
repl_user = repl
repl_password = root
```

8. 主库

```
[server1]
hostname = 192.168.1.56
port = 3306
```

9. 从库 1

```
[server2]
candidate_master = 1
check_repl_delay = 0
hostname = 192.168.1.57
port = 3306
```

10. 从库 2

```
[server3]
hostname = 192.168.1.58
port = 3306
EOF
```

12.8 创建脚本

12.8.1 自动切换脚本

自动切换脚本:master_ip_failover

```
[root@MHA-Monitor ~]# mkdir -p /usr/local/mha/scripts/
[root@MHA-Monitor ~]# vi /usr/local/mha/scripts/master_ip_failover
usr/bin/env perl
use strict
use warnings FATAL => 'all'
use Getopt::Long
my (
  $command, $ssh_user, $orig_master_host, $orig_master_ip
  $orig_master_port, $new_master_host, $new_master_ip, $new_master_port
);
```

```perl
vip 设置
my $vip = '192.168.1.54/24'
网卡名称设置
my $mhadev = 'ens33'
my $key = '1';
#my $ssh_start_vip = '/sbin/ip addr add $vip dev $mhadev'
my $ssh_start_vip = '/sbin/ifconfig $mhadev:$key $vip'
#my $ssh_stop_vip = '/sbin/ifconfig $mhadev:$key down'
my $ssh_stop_vip = '/sbin/ip addr del $vip dev $mhadev'
GetOptions(
'command=s' => \$command,
'ssh_user=s' => \$ssh_user,
'orig_master_host=s' => \$orig_master_host,
'orig_master_ip=s' => \$orig_master_ip,
'orig_master_port=i' => \$orig_master_port,
'new_master_host=s' => \$new_master_host,
'new_master_ip=s' => \$new_master_ip,
'new_master_port=i' => \$new_master_port
)
exit &main()
sub main {
print "\n\nIN SCRIPT TEST ==== $ssh_stop_vip == $ssh_start_vip === \n\n"
if ( $command eq 'stop' || $command eq 'stopssh' ) {
my $exit_code = 1
eval {
print "Disabling the VIP on old master: $orig_master_host \n"
&stop_vip()
$exit_code = 0
}
if ($@) {
warn "Got Error: $@\n"
exit $exit_code
}
exit $exit_code
}
elsif ( $command eq 'start' ) {
```

```perl
my $exit_code = 10
eval {
print 'Enabling the VIP - $vip on the new master - $new_master_host \n'
&start_vip()
$exit_code = 0
};
if ( $@ ) {
warn $@
exit $exit_code
}
exit $exit_code;
}
elsif ( $command eq 'status' ) {
print 'Checking the Status of the script.. OK \n'
exit 0
}
else {
&usage()
exit 1
}
}
sub start_vip() {
`ssh $ssh_user\@$new_master_host \' $ssh_start_vip \'`
}
sub stop_vip() {
return 0 unless ( $ssh_user)
`ssh $ssh_user\@$orig_master_host \' $ssh_stop_vip \'`
}
sub usage {
print
'Usage: master_ip_failover --command = start|stop|stopssh|status --orig_master_host = host --orig_master_ip = ip --orig_master_port = port --new_master_host = host --new_master_ip = ip --new_master_port = port\\n'
}
```

[root@MHA-Monitor ~]# chmod +x /usr/local/mha/scripts/master_ip_failover

master_ip_failover 测试脚本(Monitor 上执行)

注意:测试前要确定有ip命令 yum-y install initscripts,集群里都要装。

#停 VIP

```
/usr/local/mha/scripts/master_ip_failover --command=stop \
--ssh_user=root --orig_master_host=192.168.1.56 --orig_master_port=3306 \
--new_master_host=192.168.1.57 --new_master_port=3306
```

#启动 VIP

```
/usr/local/mha/scripts/master_ip_failover --command=start \
--ssh_user=root --orig_master_host=192.168.1.56 --orig_master_port=3306 \
--new_master_host=192.168.1.57 --new_master_port=3306
```

此脚本主要用于切换VIP。

12.8.2 手动切换脚本

```
[root@MHA-Monitor ~]# find / -name master_ip_online_change
/root/mha4mysql-manager-0.58/samples/scripts/master_ip_online_change
[root@MHA-Monitor /]# cp
/root/mha4mysql-manager-0.58/samples/scripts/master_ip_online_change
/usr/local/mha/scripts/
[root@MHA-Monitor ~]# chmod +x /usr/local/mha/scripts/master_ip_online_change
```

12.9 MHA 服务启动

12.9.1 检查 SSH 情况

```
[root@MHA-Monitor ~]# masterha_check_ssh --conf=/etc/mha/mha.cnf
```

报错处理:缺少 perl-Mail-Sender 和 perl-Log-Dispatch 这两个语言环境。

```
[root@MHA-Monitor ~]# yum install perl-Mail-Sender
[root@MHA-Monitor ~]# yum install perl-Log-Dispatch
[root@MHA-Monitor ~]# yum install epel-release --nogpgcheck -y
```

安装以下依赖即可。

```
yum install -y perl-DBD-MySQL perl-Config-Tiny perl-Log-Dispatch
perl-Parallel-ForkManager perl-YAML-Tiny perl-PAR-Dist perl-Module-ScanDeps
perl-Module-CoreList perl-Module-Build perl-CPAN perl-CPANPLUS
```

perl-File-Remove

perl-Module-Install

结果"All SSH connection tests passed successfully."表示 MHA 的 3 个数据节点之间的 SSH 是正常的。

12.9.2 检查复制情况

[root@MHA-Monitor ~]# masterha_check_repl --conf=/etc/mha/mha.cnf

"MySQL Replication Health is OK."表示 1 主 2 从的架构目前是正常的。

12.9.3 检查 MHA 状态

[root@MHA-Monitor /]# masterha_check_status --conf=/etc/mha/mha.cnf

mha is stopped(2:NOT_RUNNING).

注意:如果正常,会显示"PING_OK",否则会显示"NOT_RUNNING",这代表 MHA 监控没有开启。

12.9.4 启动 MHA Manager

[root@MHA-Monitor /]# nohup masterha_manager --conf=/etc/mha/mha.cnf < /dev/null > /usr/local/mha/manager_start.log 2>&1 &

此处多启动几次,检查结果显示"PING_OK",表示 MHA 监控软件已经启动了,主库为 192.168.1.56。

注意:一旦自动 failover 发生,mha manager 就停止监控了,需要手动再次开启。

此时一定要确定 VIP 是否正常。

12.9.5 关闭 MHA-manager

[root@MHA-Monitor /]# masterha_stop --conf=/etc/mha/mha.cnf

12.10 MHA 故障转移

12.10.1 模拟主库 Down 机

[root@MHA-Master ~]# ifconfig

ens33: flags=4163<UP,BROADCAST,RUNNING,MULTICAST> mtu 1500

    inet 192.168.1.56 netmask 255.255.255.0 broadcast 192.168.1.255

    inet6 fe80::6e36:de88:3e08:2788 prefixlen 64 scopeid 0x20<link>

ether 00:50:56:24:4c:9c txqueuelen 1000 (Ethernet)
RX packets 3307   bytes 370292 (361.6 KiB)
RX errors 0   dropped 0   overruns 0   frame 0
TX packets 2643   bytes 586340 (572.5 KiB)
TX errors 0   dropped 0 overruns 0   carrier 0   collisions 0

ens33:1: flags = 4163<UP,BROADCAST,RUNNING,MULTICAST>   mtu 1500
inet 192.168.1.54   netmask 255.255.255.0   broadcast 192.168.1.255
ether 00:50:56:24:4c:9c txqueuelen 1000 (Ethernet)

lo: flags = 73<UP,LOOPBACK,RUNNING>   mtu 65536
inet 127.0.0.1   netmask 255.0.0.0
inet6 ::1   prefixlen 128   scopeid 0x10<host>
loop txqueuelen 1   (Local Loopback)
RX packets 36   bytes 2932 (2.8 KiB)
RX errors 0   dropped 0   overruns 0   frame 0
TX packets 36   bytes 2932 (2.8 KiB)
TX errors 0   dropped 0 overruns 0   carrier 0   collisions 0

此时主库服务器掉电,关机了。

[root@MHA-Master ~]# shutdown now -h

12.10.2 查看从库 MHA-Slave1 是否为主库,此时查看 VIP 已经漂移过来

[root@MHA-Slave1 ~]# ifconfig
ens33: flags = 4163<UP,BROADCAST,RUNNING,MULTICAST>   mtu 1500
inet 192.168.1.57   netmask 255.255.255.0   broadcast 192.168.1.255
inet6 fe80::6e36:de88:3e08:2788   prefixlen 64   scopeid 0x20<link>
inet6 fe80::efe1:5d84:7c23:aa68   prefixlen 64   scopeid 0x20<link>
inet6 fe80::1a36:43a5:1f86:59c7   prefixlen 64   scopeid 0x20<link>
ether 00:50:56:36:f8:7b   txqueuelen 1000   (Ethernet)
RX packets 2804   bytes 331816 (324.0 KiB)
RX errors 0   dropped 0   overruns 0   frame 0
TX packets 2556   bytes 938972 (916.9 KiB)
TX errors 0   dropped 0 overruns 0   carrier 0   collisions 0

ens33:1: flags = 4163<UP,BROADCAST,RUNNING,MULTICAST>   mtu 1500
inet 192.168.1.54   netmask 255.255.255.0   broadcast 192.168.1.255
ether 00:50:56:36:f8:7b   txqueuelen 1000   (Ethernet)

lo: flags = 73<UP,LOOPBACK,RUNNING>   mtu 65536

inet 127.0.0.1 netmask 255.0.0.0

inet6 ::1 prefixlen 128 scopeid 0x10<host>

loop txqueuelen 1 (Local Loopback)

RX packets 36 bytes 2932 (2.8 KiB)

RX errors 0 dropped 0 overruns 0 frame 0

TX packets 36 bytes 2932 (2.8 KiB)

TX errors 0 dropped 0 overruns 0 carrier 0 collisions 0

此时发现,MHA-Slave1 做为主库,MHA-Slave2 做为从库。

```
mysql> show master status \G;
*************************** 1. row ***************************
            File: MHA-Slave1-bin.000001
        Position: 155
    Binlog_Do_DB:
Binlog_Ignore_DB: mysql,information_schema,performance_schema,sys
Executed_Gtid_Set: 0aff2757-44b5-11ed-a3bc-005056244c9c:1-11
1 row in set (0.00 sec)
```

第 13 章 高可用之多源复制

13.1 多源复制简介

所谓多源复制(Multi-source Replication),就是多台主库的数据同步到一台从库服务器上,从库创建通往每个主库的管道,从 Mysql5.7 版本开始支持多主一从的复制方式,多源复制的出现对于分库分表的业务提供了极大的便利,搭建过程支持 GTID 复制模式和 binlog+position 方式复制。与普通的复制相比,多源复制就是使用 FOR CHANNEL 进行了分离,要开启多源复制功能必须需要在从库上设置 master-info-repository 和 relay-log-info-repository 这两个参数。这两个参数是用来存储同步信息的,可以设置的值为 FILE 和 TABLE。

在部署多主一从之前,我们先了解下多源复制的好处:

(1)可以集中备份,在从库上备份,不会影响线上的数据正常运行。
(2)节约购买从库服务器的成本,只需要一个从库服务器即可。
(3)数据汇总在一起,方便后期做数据统计,可用于数仓。
(4)减轻 DBA 维护工作量,整合数据资源。

13.2 多源复制部署

13.2.1 主从规划

本次我们以 GTID 复制方式介绍多主一从的搭建方式,表 13-1 所列为主从规划。

表 13-1 多源复制主从规划

角色	Serverid	主机 IP	主机名
Master	80193265	192.168.1.33	Master1
Master	80193266	192.168.1.44	Master2
Master	80193267	192.168.1.55	Master3
Salve	80193268	192.168.1.66	Slave

13.2.2 参数文件配置

1. Master1 主库配置

cat > /etc/my.cnf << "EOF"
[mysqld]
user = mysql
port = 3306
basedir = /usr/local/mysql80/mysql8019
datadir = /usr/local/mysql80/mysql8019/data
character_set_server = utf8mb4
secure_file_priv =
server-id = 80193265
log-bin =
binlog_format = row
expire_logs_days = 30
max_binlog_size = 100M
gtid-mode = ON
enforce-gtid-consistency = on
skip_name_resolve
report_host = 192.168.1.33
EOF

2. Master2 主库配置

cat > /etc/my.cnf << "EOF"
[mysqld]
user = mysql
port = 3306
basedir = /usr/local/mysql80/mysql8019
datadir = /usr/local/mysql80/mysql8019/data
character_set_server = utf8mb4
secure_file_priv =
server-id = 80193266
log-bin =
binlog_format = row
expire_logs_days = 30
max_binlog_size = 100M

```
gtid-mode=ON
enforce-gtid-consistency=on
skip_name_resolve
report_host=192.168.1.44
EOF
```

3. Master3 主库配置

```
cat > /etc/my.cnf << "EOF"
[mysqld]
user=mysql
port=3306
basedir=/usr/local/mysql80/mysql8019
datadir=/usr/local/mysql80/mysql8019/data
character_set_server=utf8mb4
secure_file_priv=
server-id = 80193267
log-bin =
binlog_format=row
expire_logs_days = 30
max_binlog_size = 100M
gtid-mode=ON
enforce-gtid-consistency=on
skip_name_resolve
report_host=192.168.1.55
EOF
```

4. Slave 从库配置

```
cat > /etc/my.cnf << "EOF"
[mysqld]
user=mysql
port=3306
basedir=/usr/local/mysql80/mysql8019
datadir=/usr/local/mysql80/mysql8019/data
character_set_server=utf8mb4
secure_file_priv=
server-id = 80193268
log-bin =
```

```
binlog_format = row
expire_logs_days = 30
max_binlog_size = 100M
gtid-mode = ON
enforce-gtid-consistency = on
skip_name_resolve
report_host = 192.168.1.66
EOF
```

13.2.3 重启 MySQL

```
mysqladmin -uroot -p shutdown
mysqld_safe &
mysql -uroot -p
```

13.2.4 主库配置

在 3 台主机 Master1、Master2、Master3 做相同操作,如图 13-1 所示。

```
mysql> create user repl@'%' identified with mysql_native_password by 'root'
mysql> grant all on *.* to repl@'%' with grant optio
mysql> flush privileges
mysql> select user,host,grant_priv,password_last_changed authentication_string from mysql.user

mysql> show master status \G
mysql> show slave hosts
mysql> select @@hostname,@@server_id,@@server_uuid
```

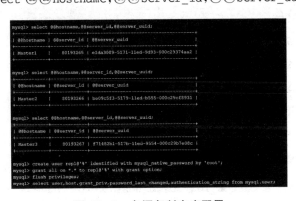

图 13-1 多源复制主库配置

13.2.5 从库配置

1. 登录从库

详细的操作过程如图 13-2 所示，使用 FOR CHANNEL 进行了分离。

mysql -uroot -p -h192.168.1.66

2. 从库连接主库 Master1

change master to
master_host = '192.168.1.33'
master_port = 3306,master_user = 'repl'
master_password = 'root'
master_auto_position = 1 FOR CHANNEL 'Master1'

3. 从库连接主库 Master2

change master to
master_host = '192.168.1.44'
master_port = 3306,master_user = 'repl'
master_password = 'root'
master_auto_position = 1 FOR CHANNEL 'Master2'

4. 从库连接主库 Master3

change master to
master_host = '192.168.1.55'
master_port = 3306,master_user = 'repl'
master_password = 'root'
master_auto_position = 1 FOR CHANNEL 'Master3'

5. 启动 SLAVE

--启动所有 SLAVE
mysql> START SLAVE

--也可以单独启动需要同步的通道
START SLAVE FOR CHANNEL 'master2'
START SLAVE FOR CHANNEL 'master3'

6. 查询主库信息

mysql> select a.master_log_pos,a.host,a.user_name,a.user_password,
a.port,a.uuid,a.channel_name from mysql.slave_master_info a

7. 已经执行过事务信息

mysql> select * from mysql.gtid_executed

8. 查询主从应用

SHOW SLAVE STATUS FOR CHANNEL 'master1'\G

SHOW SLAVE STATUS FOR CHANNEL 'master2'\G

SHOW SLAVE STATUS FOR CHANNEL 'master3'\G

```
mysql> change master to
    -> master_host='192.168.1.33',
    -> master_port=3306,master_user='rep1',
    -> master_password='root',
    -> master_auto_position=1 FOR CHANNEL 'Master1';
Query OK, 0 rows affected, 2 warnings (0.03 sec)

mysql> change master to
    -> master_host='192.168.1.44',
    -> master_port=3306,master_user='rep1',
    -> master_password='root',
    -> master_auto_position=1 FOR CHANNEL 'Master2';
Query OK, 0 rows affected, 2 warnings (0.02 sec)

mysql> change master to
    -> master_host='192.168.1.55',
    -> master_port=3306,master_user='rep1',
    -> master_password='root',
    -> master_auto_position=1 FOR CHANNEL 'Master3';
Query OK, 0 rows affected, 2 warnings (0.02 sec)

mysql> START SLAVE;
Query OK, 0 rows affected (0.01 sec)

mysql> select a.master_log_pos,a.host,a.user_name,a.user_password,a.port,a.uuid,a.channel_name
    -> from mysql.slave_master_info a;
+----------------+--------------+-----------+---------------+------+--------------------------------------+--------------+
| master_log_pos | host         | user_name | user_password | port | uuid                                 | channel_name |
+----------------+--------------+-----------+---------------+------+--------------------------------------+--------------+
|            155 | 192.168.1.33 | rep1      | root          | 3306 | e1da3089-5171-11ed-9d93-000c29374aa2 | master1      |
|            155 | 192.168.1.44 | rep1      | root          | 3306 | be09c5f3-5179-11ed-b555-000c29cf8931 | master2      |
|            155 | 192.168.1.55 | rep1      | root          | 3306 | f71482b1-517b-11ed-9554-000c29b7e08c | master3      |
+----------------+--------------+-----------+---------------+------+--------------------------------------+--------------+
3 rows in set (0.00 sec)
```

图 13-2 使用 FOR CHANNEL 进行了分离

13.2.6 测试多源复制

1. 主库 Master1 测试

mysql -uroot -p -h192.168.1.33

mysql> create database master1

mysql> use master1

mysql> CREATE TABLE `test1` (`id` int(11) DEFAULT NULL,`count` int(11) DEFAULT NULL)

mysql> insert into test1 values(1,1)

mysql> SELECT * FROM master1.test1

2. 主库 Master2 测试

mysql -uroot -p -h192.168.1.44

mysql> create database master2

mysql> use master2

```
mysql> CREATE TABLE `test2` (`id` int(11) DEFAULT NULL,`count` int(11) DEFAULT NULL)
mysql> insert into test2 values(2,2)
mysql> SELECT * FROM master2.test2
```

3. 主库 Master3 测试

```
mysql -uroot -p -h192.168.1.55
mysql> create database master3
mysql> use master3
mysql> CREATE TABLE `test3` (`id` int(11) DEFAULT NULL,`count` int(11) DEFAULT NULL)
mysql> insert into test3 values(3,3)
mysql> SELECT * FROM master3.test3
```

4. 从库 Slave 查询

从库已应用如图 13-3 所示。

```
mysql -uroot -p -h192.168.1.66
mysql> show databases
mysql> SELECT * FROM master1.test1
mysql> SELECT * FROM master2.test2
mysql> SELECT * FROM master3.test3
```

```
mysql> SELECT * FROM master1.test1;
+----+-------+
| id | count |
+----+-------+
|  1 |     1 |
+----+-------+
1 row in set (0.00 sec)
mysql> SELECT * FROM master2.test2;
+----+-------+
| id | count |
+----+-------+
|  2 |     2 |
+----+-------+
1 row in set (0.00 sec)
mysql> SELECT * FROM master3.test3;
+----+-------+
| id | count |
+----+-------+
|  3 |     3 |
+----+-------+
1 row in set (0.00 sec)
mysql> show databases;
+--------------------+
| Database           |
+--------------------+
| information_schema |
| master1            |
| master2            |
| master3            |
| mysql              |
| performance_schema |
| sys                |
+--------------------+
```

图 13-3　从库 Slave 应用

注意：

(1)初次配置耗时较长，需要将各 master 的数据 dump 下来，再 source 到 slave 上。

(2)需要考虑各 master 数据增长频率，slave 的数据增长频率是这些数据的总和。如果太高，会导致大量的磁盘 IO，造成数据更新延迟，严重时是会影响正常的查询。

(3)如果多个主数据库实例中存在同名的库，则同名库的表都会放到一个库中。

(4)如果同名库中的表名相同且结构相同，则数据会合并到一起；如果结构不同，则先建的有效。

第 14 章 高可用之 MGR 架构

14.1 MGR 简介

MySQL Group Replication(简称 MGR)是一款高可用与高扩展的解决方案,它提供了高可用、高扩展、高可靠的 MySQL 集群服务。既可以很好地保证数据一致性又可以自动切换,具备故障检查功能、支持多节点写入,以插件形式提供,实现分布式下数据的最终一致性,MGR 是从 MySQL 5.7.17 版本推出的。

MGR 是 MySQL 自带的一个插件,可以灵活部署,MySQL MGR 集群是多个 MySQL Server 节点共同组成的分布式集群,每个 Server 都有完整的副本,它是基于 ROW 格式的二进制日志文件和 GTID 特性。适用于金融交易、重要数据存储、对主从一致性要求高的场景,读多写少的应用场景,如互联网电商。

14.2 MGR 特点

14.2.1 强一致性

基于原生复制及 paxos 协议的组复制技术,并以插件方式提供,提供一致性数据安全。

14.2.2 高容错性

只要不是大多数节点环境坏掉就可以继续工作,有自动检查机制,当不同节点产生资源争用冲突时,不会出现错误,按照先到者优先原则进行处理,并内置自动化脑裂防护机制。

14.2.3 高扩展性

节点的新增和移除都是自动的,新节点加入后,会自动从其他节点上同步状态,直到新节点和其他节点保持一致,如果某节点被移除了,其他节点自动更新组信息,自动维护新的组信息。

14.2.4 高灵活性

有单主模式和多主模式,单主模式下会自动选主,所有的更新操作都是在主上进行;多主模式下,所有 server 都可以同时处理更新操作。

14.3 MGR 部署

接下来我们按照图 14-1 基于 Linux8 环境 MGR 集群部署,MySQL 版本为 8.0.19。

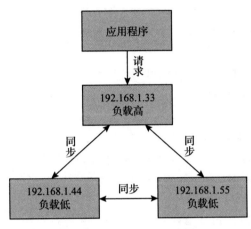

图 14-1 MGR 集群架构

14.3.1 参数文件配置

1. Master1 配置

cat > /etc/my.cnf <<"EOF"
[mysqld]
user = mysql
port = 3306
character_set_server = utf8mb4
secure_file_priv =
server-id = 801933063
log_timestamps = SYSTEM
log-bin =
binlog_format = row
binlog_checksum = NONE
skip-name-resolve
log_slave_updates = 1
gtid-mode = ON

```
enforce-gtid-consistency=on
default_authentication_plugin=mysql_native_password
max_allowed_packet = 500M
master_info_repository=TABLE
relay_log_info_repository=TABLE
relay_log=mgr01-relay-bin
transaction_write_set_extraction=XXHASH64
loose-group_replication_group_name="bbbbbbbb-bbbb-bbbb-bbbb-bbbbbbbbbbbb"
loose-group_replication_start_on_boot=OFF
loose-group_replication_local_address="192.168.1.33:33061"
loose-group_replication_group_seeds="192.168.1.33:33061,192.168.1.44:33061,192.168.1.55:33061"
loose-group_replication_bootstrap_group=OFF
loose-group_replication_ip_whitelist="192.168.1.33,192.168.1.44,192.168.1.55"
report_host=192.168.1.33
report_port=3306
EOF
```

2. Master2 配置

```
cat > /etc/my.cnf <<"EOF"
[mysqld]
user=mysql
port=3306
character_set_server=utf8mb4
secure_file_priv=
server-id = 801933064
log_timestamps = SYSTEM
log-bin=
binlog_format=row
binlog_checksum=NONE
log_slave_updates = 1
gtid-mode=ON
enforce-gtid-consistency=ON
skip_name_resolve
```

```
default_authentication_plugin = mysql_native_password
max_allowed_packet = 500M
master_info_repository = TABLE
relay_log_info_repository = TABLE
relay_log = mgr02-relay-bin
transaction_write_set_extraction = XXHASH64
loose-group_replication_group_name = "bbbbbbbb-bbbb-bbbb-bbbb-bbbbbbbbbbbb"
loose-group_replication_start_on_boot = OFF
loose-group_replication_local_address = "192.168.1.44:33061"
loose-group_replication_group_seeds = "192.168.1.33:33061,192.168.1.44:33061,192.168.1.55:33061"
loose-group_replication_bootstrap_group = OFF
loose-group_replication_ip_whitelist = "192.168.1.33,192.168.1.44,192.168.1.55"
report_host = 192.168.1.44
report_port = 3306
EOF
```

3. Master3 配置

```
cat > /etc/my.cnf <<"EOF"
[mysqld]
user = mysql
port = 3306
character_set_server = utf8mb4
secure_file_priv =
server-id = 801933065
log_timestamps = SYSTEM
log-bin =
binlog_format = row
binlog_checksum = NONE
log_slave_updates = 1
gtid-mode = ON
enforce-gtid-consistency = ON
skip_name_resolve
default_authentication_plugin = mysql_native_password
```

```
max_allowed_packet = 500M
master_info_repository = TABLE
relay_log_info_repository = TABLE
relay_log = mgr03-relay-bin
transaction_write_set_extraction = XXHASH64
loose-group_replication_group_name = "bbbbbbbb-bbbb-bbbb-bbbb-bbbbbbbbbbbb"
loose-group_replication_start_on_boot = OFF
loose-group_replication_local_address = "192.168.1.55:33061"
loose-group_replication_group_seeds = "192.168.1.33:33061,192.168.1.44:33061,192.168.1.55:33061"
loose-group_replication_bootstrap_group = OFF
loose-group_replication_ip_whitelist = "192.168.1.33,192.168.1.44,192.168.1.55"
report_host = 192.168.1.55
report_port = 3306
EOF
```

14.3.2 安装 MGR 插件

先重启 MySQ,在登录,在 MGR 集群 3 个节点均操作:

```
[root@Master1 ~]# mysql -uroot -p
mysql> INSTALL PLUGIN group_replication SONAME 'group_replication.so'
mysql> show plugins
```

14.3.3 设置复制账号

```
mysql> SET SQL_LOG_BIN = 0
mysql> create user repl@'%' identified with mysql_native_password by 'jem'
mysql> GRANT REPLICATION SLAVE ON *.* TO repl@'%'
mysql> FLUSH PRIVILEGES
mysql> SET SQL_LOG_BIN = 1
```

注意:如果想在主库上执行一些操作,但不复制到 slave 库上,可以通过修改参数 sql_log_bin 来实现。

8.0.23 版本以下:

```
CHANGE MASTER TO master_user = 'repl',master_password = 'jem' FOR CHANNEL 'group_replication_recovery'
```

8.0.23 版本及以上：

CHANGE REPLICATION SOURCE to SOURCE_USER = 'repl', SOURCE_PASSWORD = 'jem' FOR CHANNEL 'group_replication_recovery'

14.4　MGR 单主模式

14.4.1　启动主节点 MGR

--在主库(192.168.1.33)上执行

mysql> SET GLOBAL group_replication_bootstrap_group = ON

mysql> START GROUP_REPLICATION

mysql> SET GLOBAL group_replication_bootstrap_group = OFF

第一次启动组的过程称为引导。使用 group_replication_bootstrap_group 系统变量来引导一个组。引导程序只能由单个 MySQL Server（这里指的是引导组的 MySQL Server）执行一次。这就是为什么 group_replication_bootstrap_group 系统变量不在 my.cnf 配置文件中持久化的原因。

如果将其保存在配置文件中（group_replication_bootstrap_group＝ON），则在重新启动组成员时，组中的所有成员都将会尝试引导组，而这些组的名称都是相同的，这将导致组产生脑裂。因此，为了安全地引导组，需要在第一个 MySQL Server 启动完成之后，登录到数据库中，手工执行如下语句完成组的引导（该参数也可用于重新引导组，先设置为 OFF，再设置为 ON），单主模式如图 14-2 所示。

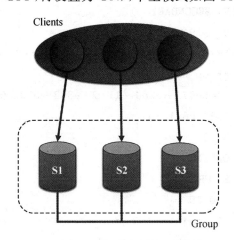

图 14-2　MGR 单主集群架构

mysql> SELECT * FROM performance_schema.replication_group_members\G
*************************** 1. row ***************************

```
      CHANNEL_NAME: group_replication_applier
        MEMBER_ID: 14efe973-52c4-11ed-a3b3-000c29374aa2
      MEMBER_HOST: 192.168.1.33
      MEMBER_PORT: 3306
     MEMBER_STATE: ONLINE
      MEMBER_ROLE: PRIMARY
   MEMBER_VERSION: 8.0.19
1 row in set (0.00 sec)
```

14.4.2 其他节点加入 MGR

--在从库(192.168.1.44,192.168.1.55)上执行,先结束当前应用的事务

```
mysql> reset master
mysql> START GROUP_REPLICATION
```

--查看 MGR 组信息

```
mysql> SELECT * FROM performance_schema.replication_group_members\G
*************************** 1. row ***************************
      CHANNEL_NAME: group_replication_applier
        MEMBER_ID: 020ef209-52c6-11ed-9291-000c29cf8931
      MEMBER_HOST: 192.168.1.44
      MEMBER_PORT: 3306
     MEMBER_STATE: ONLINE
   1. MEMBER_ROLE: SECONDARY
   MEMBER_VERSION: 8.0.19
*************************** 2. row ***************************
      CHANNEL_NAME: group_replication_applier
        MEMBER_ID: 14efe973-52c4-11ed-a3b3-000c29374aa2
      MEMBER_HOST: 192.168.1.33
      MEMBER_PORT: 3306
     MEMBER_STATE: ONLINE
   2. MEMBER_ROLE: PRIMARY
   MEMBER_VERSION: 8.0.19
*************************** 3. row ***************************
      CHANNEL_NAME: group_replication_applier
        MEMBER_ID: 3f3ef0e8-52cb-11ed-8707-000c29b7e08c
      MEMBER_HOST: 192.168.1.55
      MEMBER_PORT: 3306
```

```
    MEMBER_STATE: ONLINE
  3. MEMBER_ROLE: SECONDARY
MEMBER_VERSION: 8.0.19
3 rows in set (0.00 sec)
```

--查询当前模式
```
mysql> show variables like '%group_replication_single_primary_mode%'\G
*************************** 1. row ***************************
Variable_name: group_replication_single_primary_mode
        Value: ON
1 row in set (0.01 sec)
```

从上面信息可以看到,3个节点状态为online,并且主节点为192.168.1.33,只有主节点可以写入,其他节点只读,MGR单主模式搭建成功。

参数group_replication_single_primary_mode为ON,表示单主模式。

14.5 多主和单主模式切换

14.5.1 查询当前模式

```
mysql> show variables like '%group_replication_single_primary_mode%'
mysql> SELECT @@group_replication_single_primary_mode
```

参数group_replication_single_primary_mode为ON,表示单主模式

本次使用函数切换,从MySQL 8.0.13开始,可以使用函数进行在线修改MGR模式,不需要重启组复制。

14.5.2 单主切多主

MGR多主模式如图14-3所示。

--31上主库执行即可
```
mysql> select group_replication_switch_to_multi_primary_mode()
+------------------------------------------------------+
| group_replication_switch_to_multi_primary_mode()     |
+------------------------------------------------------+
| Mode switched to multi-primary successfully.         |
+------------------------------------------------------+
1 row in set (1.03 sec)

mysql> SELECT * FROM performance_schema.replication_group_members\G
```

```
*************************** 1. row ***************************
    CHANNEL_NAME: group_replication_applier
      MEMBER_ID: 020ef209-52c6-11ed-9291-000c29cf8931
    MEMBER_HOST: 192.168.1.44
    MEMBER_PORT: 3306
   MEMBER_STATE: ONLINE
 1. MEMBER_ROLE: PRIMARY
MEMBER_VERSION: 8.0.19
*************************** 2. row ***************************
    CHANNEL_NAME: group_replication_applier
      MEMBER_ID: 14efe973-52c4-11ed-a3b3-000c29374aa2
    MEMBER_HOST: 192.168.1.33
    MEMBER_PORT: 3306
   MEMBER_STATE: ONLINE
 2. MEMBER_ROLE: PRIMARY
MEMBER_VERSION: 8.0.19
*************************** 3. row ***************************
    CHANNEL_NAME: group_replication_applier
      MEMBER_ID: 3f3ef0e8-52cb-11ed-8707-000c29b7e08c
    MEMBER_HOST: 192.168.1.55
    MEMBER_PORT: 3306
   MEMBER_STATE: ONLINE
 3. MEMBER_ROLE: PRIMARY
MEMBER_VERSION: 8.0.19
3 rows in set (0.00 sec)
```

图 14-3 MGR 多主模式

从上面 14-3 信息可以看到，3 个节点状态为 online，均变成了主节点，只有主节点可以写入，其他节点只读，MGR 多主模式切换成功。

参数 group_replication_single_primary_mode 为 0，表示多主模式。

14.5.3 多主切单主

多主切单主,入参需要传入主库的 server_uuid,查询具体的 uuid,在这里我将节点 3(IP:192.168.1.55)作为主库。

```
mysql> select @@server_uuid\G
*************************** 1. row ***************************
@@server_uuid: 3f3ef0e8-52cb-11ed-8707-000c29b7e08c
mysql> select group_replication_switch_to_single_primary_mode('3f3ef0e8-52cb-11ed-8707-000c29b7e08c')
+----------------------------------------------------------------------------------+
| group_replication_switch_to_single_primary_mode('3f3ef0e8-52cb-11ed-8707-000c29b7e08c') |
+----------------------------------------------------------------------------------+
| Mode switched to single-primary successfully.                                    |
+----------------------------------------------------------------------------------+
mysql> select MEMBER_ID,MEMBER_HOST,MEMBER_PORT,MEMBER_STATE,MEMBER_ROLE from performance_schema.replication_group_members
```

MEMBER_ID	MEMBER_HOST	MEMBER_PORT	MEMBER_STATE	MEMBER_ROLE
020ef209-52c6-11ed-9291-000c29cf8931	192.168.1.44	3306	ONLINE	SECONDARY
14efe973-52c4-11ed-a3b3-000c29374aa2	192.168.1.33	3306	ONLINE	SECONDARY
3f3ef0e8-52cb-11ed-8707-000c29b7e08c	192.168.1.55	3306	ONLINE	PRIMARY

从上面信息可以看到,3 个节点状态为 online,节点 3(IP:192.168.1.55)变成了主库,其他的两个节点变成了从库。

第 15 章 双主＋Keepalived 单点故障切换

15.1 Keepalived 简介

Keepalived 官网界面如图所示。
官方网址：https://www.keepalived.org/。
官网文档：http://www.keepalived.org/documentation.html。

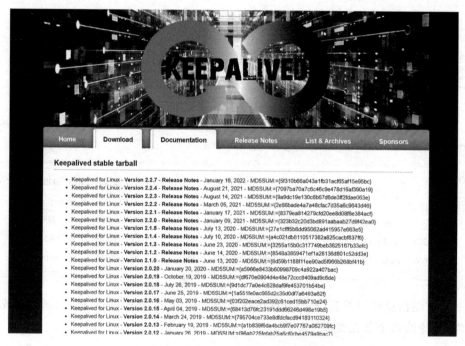

图 15-1 Keepalived 官网主界面

keepalived 是集群管理中保证集群高可用的一个服务软件，其功能类似于 heartbeat，用来防止单点故障。Keepalived 高可用故障切换，是通过 VRRP 虚拟路由器冗余协议来实现的。keepalived 主要有三个模块，分别是 core、check 和 vrrp。core 模块为 keepalived 的核心，负责主进程的启动、维护以及全局配置文件的加载和解析。check 负责健康检查，包括常见的各种检查方式。vrrp 模块是来实现 VRRP 协议的。

一般中小型公司都使用这种架构，搭建比较方便简单；可以采用主从或者主主模式，在 master 节点发生故障后，利用 keepalived 高可用机制实现快速切换到 slave

节点。原来的从库变成新的主库。

使用 Keepalived 的 HA 功能可实现 MySQL 主从复制的自动故障切换。它的工作原理是：初始将 MySQL 的主从两个主机赋予不同的优先级别，当 Keepalived 启动时，会将 VIP 绑定到高优先级的主库上。在 Keepalived 中调用自定义脚本 check_run，每分钟检查一次本机 MySQL 的服务器状态，如果 MySQL 不可用，则关掉本机的 keepalived 进程。Keepalived 每秒钟会检查一次本机的 keepalived 进程，如果进程不存在，则将 VIP 绑定到另一台机器上，如果这台机器原来是从库，则同时调用 master.sh 脚本执行从库切换为主库的操作。

15.2 双主复制要点

与主从复制相比，双主复制需要注意以下三个参数的设置：

log_slave_updates：要设置为 true，将复制事件写入本机 binlog。一台服务器既做主库又做从库时此选项必须要开启。

auto_increment_offset 和 auto_increment_increment：为避免自增列冲突，需要设置这两个参数，例如在双主复制中，可以配置如下：

\# masterA 自增长 ID
auto_increment_offset = 1
auto_increment_increment = 2 \# 奇数 ID
\# masterB 自增加 ID
auto_increment_offset = 2
auto_increment_increment = 2 \# 偶数 ID

其中，auto_increment_increment 是自增的步长，value 为 1 代表每次 1，auto_increment_offset 是自增的偏移量，也就是自增开始，value 为 1 代表从 1 开始增加。我们通过以上方式，做一个自增的错位，同步就 OK。

15.3 双主部署

本次将基于 Docker 环境，部署 MySQL8.0 GTID 模式的双主，硬件规划如表 15-1 列。

表 15-1 双主＋Keepalived 单点故障切换硬件规划

角色	Serverid	主机 IP	主机名
Master	330631	192.168.68.31	MMK-JEM-Master1-ip31
Master	330632	192.168.68.32	MMK-JEM-Master2-ip32

在部署之前我们先通过以下命令创建私有的网络：

```
docker network create --subnet=192.168.68.0/16 mhajem
docker network inspect mhajem
```

15.3.1 创建服务器

```
docker run -d --name MMK-JEM-Master1-ip31 \
-h MMK-JEM-Master1-ip31 \
--network mhajem --ip 192.168.68.31 --privileged=true \
-p 33461:3306 -p 2241:22 \
centos:8.2 init
```

```
docker run -d --name MMK-JEM-Master2-ip32 \
-h MMK-JEM-Master2-ip32 \
--network mhajem --ip 192.168.68.32 --privileged=true \
-p 33462:3306 -p 2242:22 \
centos:8.2 init
```

说明：centos:8.2 为下载好的操作系统镜像，以上相当于申请了两台服务器。

15.3.2 部署 MySQL8.0 双主

1. 二进制安装包

`mysql-8.0.19-linux-glibc2.12-x86_64.tar.xz`

2. 从宿主机复制安装包到容器

```
docker cp mysql-8.0.19-linux-glibc2.12-x86_64.tar.xz MMK-JEM-Master1-ip31:/
```

```
docker cp mysql-8.0.19-linux-glibc2.12-x86_64.tar.xz MMK-JEM-Master2-ip32:/
```

3. 分别进入容器(操作系统)

```
docker exec -it MMK-JEM-Master1-ip31 /bin/bash
```

```
docker exec -it MMK-JEM-Master2-ip32 /bin/bash
```

4. 用户及组添加

```
groupadd mysql
useradd -r -g mysql mysql
```

5. 解压缩，创建软链接

```
tar -xf mysql-8.0.19-linux-glibc2.12-x86_64.tar.xz -C /usr/local/
ln -s /usr/local/mysql-8.0.19-linux-glibc2.12-x86_64 /usr/local/mysql
ln -s /usr/local/mysql/mysql /usr/bin/mysql
```

6. 环境变量及授权

```
echo "export PATH=$PATH:/usr/local/mysql/bin" >> /etc/bashrc
source /etc/bashrc
chown -R mysql.mysql /usr/local/mysql-8.0.19-linux-glibc2.12-x86_64
```

7. 在线 yum 配置，加入以下链接

```
cd /etc/yum.repos.d
vi CentOS-Base.repo
baseurl=https://vault.centos.org/centos/$releasever/BaseOS/$basearch/os/

vi CentOS-AppStream.repo
baseurl=https://vault.centos.org/centos/$releasever/AppStream/$basearch/os/
```

8. 安装的依赖包

```
yum install -y net-tools
yum install -y libtinfo*
yum -y install numactl
yum -y install libaio*
yum -y install perl perl-devel
yum -y install autoconf
yum -y install numactl.x86_64
```

9. 配置参数文件

```
cat>/etc/my.cnf<<"EOF"
[mysqld]
basedir=/usr/local/mysql
datadir=/usr/local/mysql/data
user=mysql
port=3306
character_set_server=utf8mb4
```

```
secure_file_priv =
server-id=330631
log-bin=
binlog_format=STATEMENT
binlog-ignore-db=mysql
binlog-ignore-db=information_schema
binlog-ignore-db=performance_schema
binlog-ignore-db=sys
replicate_ignore_db=information_schema
replicate_ignore_db=performance_schema
replicate_ignore_db=mysql
replicate_ignore_db=sys
log-slave-updates
auto-increment-increment=2
auto-increment-offset=1
skip_name_resolve
log_timestamps=SYSTEM
default-time-zone='+8:00'
gtid-mode=ON
enforce-gtid-consistency=on
EOF
```

注意：以上为主库 M1 的参数文件。

```
cat>/etc/my.cnf<<"EOF"
[mysqld]
basedir=/usr/local/mysql
datadir=/usr/local/mysql/data
user=mysql
port=3306
character_set_server=utf8mb4
secure_file_priv=
server-id=330632
log-bin=
binlog_format=STATEMENT
binlog-ignore-db=mysql
binlog-ignore-db=information_schema
```

```
binlog-ignore-db=performance_schema
binlog-ignore-db=sys
replicate_ignore_db=information_schema
replicate_ignore_db=performance_schema
replicate_ignore_db=mysql
replicate_ignore_db=sys
log-slave-updates
auto-increment-increment=2
auto-increment-offset=2
skip_name_resolve
log_timestamps=SYSTEM
default-time-zone='+8:00'
gtid-mode=ON
enforce-gtid-consistency=ON
EOF
```

注意：以上为主库 M2 的参数文件。

10. 修改本地用户密码

mysql> alter user root@'localhost' identified with mysql_native_password by 'root';

mysql> flush privileges;

11. 创建远程用户 root

mysql> create user root@'%' identified with mysql_native_password by 'root';

mysql> grant all on *.* to root@'%' with grant option;

mysql> flush privileges;

12. 创建双主复制用户

mysql> create user repl@'%' identified with mysql_native_password by 'root';

mysql> grant all on *.* to repl@'%' with grant option;

mysql> flush privileges;

说明：请在两个主库上分别创建。

13. 双主配置

——M2 主库 mysql 窗口执行以下命令

```
change master to
master_host = '192.168.68.31'
master_port = 3306
master_user = 'repl'
master_password = 'root'
master_auto_position = 1

start slave
show slave status\G
```

--M1 主库 mysql 窗口执行以下命令
```
change masterto
master_host = '192.168.68.32'
master_port = 3306
master_user = 'repl'
master_password = 'root'
master_auto_position = 1

start slave
show slave status\G
```

注意：start slave 命令可以启动从库复制，双主模式，每台服务器即是主库也是从库。

14. 测试同步的数据

--主库 M1
```
create database jemdb
use jemdb
create table jemdb.mytb1(id int,name varchar(30))
SET SESSION binlog_format = 'STATEMENT'
insert into jemdb.mytb1 values(2,@@hostname)
select * from jemdb.mytb1
```
--主库 M2
```
create database jemdb2
use jemdb2
create table jemdb2.mytb2(id int,name varchar(30))
SET SESSION binlog_format = 'STATEMENT'
insert into jemdb2.mytb2 values(2,@@hostname)
select * from jemdb2.mytb2
```

说明:通过数据的测试,双主模式下数据实时同步。

15.3.3 配置 keepalived

1. 安装 keepalived

```
yum install -y keepalived
```

2. keepalived.conf 配置

主库 M1 和主库 M2 两台服务器同样进行如下操作:

```
cat > /etc/keepalived/keepalived.conf << "EOF"
! Configuration File for keepalived
global_defs {
router_id MySQL-MM-HA
}
#检测 mysql 服务是否在运行。有很多方式,比如进程,用脚本检测等
vrrp_script chk_mysql_port {
#这里通过脚本监测
#脚本执行间隔,每2s 检测一次
script "/etc/keepalived/chk_mysql.sh" interval 2
#脚本结果导致的优先级变更,检测失败(脚本返回非 0)则优先级 -5
weight -5
#检测连续 2 次失败才算确定是真失败,会用 weight 减少优先级(1-255 之间)
fall 2
#检测 1 次成功就算成功。但不修改优先级
rise 1
}
vrrp_instance VI_1 {
#建议设置为 BACKUP
state BACKUP
#指定虚拟 ip 的网卡接口
interface eth1
#路由器标识,MASTER 和 BACKUP 必须是一致的才能保证是同一套系统
virtual_router_id 51
#定义优先级,数字越大,优先级越高
priority 100
advert_int 1
authentication {
```

```
    auth_type PASS
    auth_pass 1111
    }
    virtual_ipaddress {
    192.168.68.36
    }
    track_script {
    chk_mysql_port
    }
    }
    EOF
```

--脚本监测

```
cat > /etc/keepalived/chk_mysql.sh << "EOF"
#!/bin/bash
counter=$(netstat -na|grep "LISTEN"|grep "3306"|wc -l)
if [ "${counter}" -eq 0 ]; then
    systemctl stop keepalived
fi
EOF
```

--配置文件授权

```
chmod 755 /etc/keepalived/chk_mysql.sh
systemctl start keepalived
systemctl enable keepalived
```

15.3.4 keepalived 故障模拟

```
mysql -uroot -proot -h192.168.68.36 -P3306
select @@server_id,@@hostname
create database jemdb3
use jemdb3
create table jemdb3.mytb1(id int,name varchar(30))
SET session binlog_format = 'STATEMENT'
insert into jemdb3.mytb1 values(1,@@hostname)
select * from jemdb3.mytb1
```

注意：通过数据的测试，双主模式下 keepalived 单点故障，可不影响业务的应用，自动切换。

第 16 章 高可用之 PXC 架构

16.1 PXC 介绍

PXC 是一套 MySQL 高可用集群解决方案，与传统的基于主从复制模式的集群架构相比，PXC 最突出特点就是解决了诟病已久的数据复制延迟问题，基本上可以达到实时同步。而且节点与节点之间相互的关系是对等的。PXC 最关注的是数据的一致性，对待事务的行为时，要么在所有节点上执行，要么都不执行，它的实现机制决定了它对待一致性的行为非常严格，这也能非常完美保证 MySQL 集群的数据一致性。

Percona XtraDB Cluster 是一个完全开源的 MySQL 的高可用性解决方案。它将 Percona Server 和 Percona XtraBackup 与 Galera 库集成，以实现同步多主复制。集群由节点组成，其中每个节点包含同一组数据同步的跨节点。推荐的配置至少有 3 个 8 节点。每个节点都是常规的 MySQL 服务器实例（例如 Percona Server）。可以将现有的 MySQL 服务器实例转换为节点，并使用此节点作为基础来运行集群。还可以从集群中分离任何节点，并将其用作常规的 MySQL 服务器实例，PXC 是在存储引擎层实现的同步复制，而非异步复制，所以其数据的一致性是相当高的，架构如图 16-1 所示。

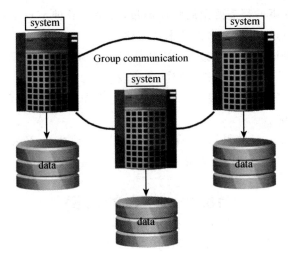

图 16-1 PXC 架构图

16.2 PXC 的优缺点

16.2.1 优点

(1)实现 mysql 数据库集群架构的高可用性和数据的强一致性。
(2)完成了真正的多节点读写的集群方案。
(3)改善了传统意义上的主从复制延迟问题,基本上达到了实时同步。
(4)新加入的节点可以自动部署,无须提供手动备份,维护起来很方便。
(5)由于是多节点写入,所以数据库故障切换很容易。

16.2.2 缺点

(1)只支持 Innodb 存储引擎。
(2)存在多节点 update 更新问题,也就是写放大问题。
(3)在线 DDL 语句,锁表问题。
(4)sst 针对新节点加入的传输代价过高的问题。
(5)所有表都要有主键。
(6)不支持 LOCK TABLE 等显式锁操作。
(7)锁冲突、死锁问题相对更多。
(8)不支持 XA:

①加入新节点时开销大,添加新节点时,必须从现有节点复制完整数据集。如果是 300 GB,则复制 300 GB。
②任何更新的事务都需要全局验证通过,才会在其他节点上执行。集群性能受限于性能最差的节点,也就是常说的木桶定律。
③因为需要保证数据的一致性,PXC 采用的实时基于存储引擎层来实现同步复制,所以在多节点并发写入时,锁冲突问题比较严重。
④存在写扩大的问题,所以节点上都会发生写操作,对于写负载过大的场景,不推荐使用 PXC 架构。
⑤只支持 innodb 存储引擎,所有表都要有主键,锁冲突、死锁问题相对更多。
⑥不支持 LOCK TABLE 等显式锁操作,不支持 XA。

16.3 PXC 的原理

PXC 集群提供了 MySQL 高可用的一种实现方法。
(1)集群是有节点组成的,推荐配置至少 3 个节点,但是也可以运行在 2 个节点上。

(2)每个节点都是普通的 mysql/percona 服务器,可以将现有的数据库服务器组成集群,反之,也可以将集群拆分成单独的服务器。

(3)每个节点都包含完整的数据副本。

PXC 集群主要由两部分组成:Percona Server with XtraDB 和 Write Set Replication patches(使用了 Galera library,一个通用的用于事务型应用的同步、多主复制插件)。

PXC 会使用大概是 4 个端口号:

① 3306 数据库对外服务的端口号。

② 4444 请求 SST,SST 指数据一个镜象传输 xtrabackup,rsync,mysqldump。

③ 4567 组成员之间进行沟通的一个端口号。

④ 4568 传输 IST 用的,相对于 SST 来说的一个增量,重启加入那一个时间有用。

PXC 的操作流程如下:

(1)首先客户端先发起一个事务,该事务先在本地执行,执行完成之后就要发起对事务的提交操作了。

(2)在提交之前需要将产生的复制写集广播出去,然后获取到一个全局的事务 ID 号,一并传送到另一个节点上面。

(3)通过合并数据之后,发现没有冲突数据,执行 apply_cd 和 commit_cb 动作,否则就需要取消此次事务的操作。

(4)当前 server 节点通过验证之后,执行提交操作,并返回 OK,如果验证没通过,则执行回滚。

(5)在生产中至少要有 3 个节点的集群环境,如果其中一个节点没有验证通过,出现了数据冲突,那么此时采取的方式就是将出现不一致的节点踢出集群环境,而且它自己会执行 shutdown 命令,自动关机。

16.4　PXC 的重要概念

16.4.1　名词介绍

(1)WS:write set 写数据集,write-sets:事务写集。

(2)IST:Incremental State Transfer 数据增量同步,节点加入集群的数据同步方式。

(3)SST:State Snapshot Transfer 全量同步,数据全量同步,节点加入集群的数据同步方式。

(4)Donor 节点:新节点加入集群时,集群会投票选出 Donor 节点提供新节点所需的数据。

首先要规范集群中节点的数量,整个集群节点数控制在最少3个、最多8个的范围内。最少3个是为了防止"脑裂"现象,因为只有在两个节点的情况下才会出现脑裂。极端的表现就是输出任何命令,返回的结果都是 unknow command。

当一个新节点要加入 PXC 集群时,需要从集群中各节点里选举一个 doner 节点作为全量数据的贡献者。PXC 有两种节点的数据传输方式,一种叫 SST 全量传输,另一种叫 IST 增量传输。SST 传输有 XtraBackup、mysqldump、rsync 三种方式,而增量传输只有 XtraBackup。一般数据量不大时可以使用 SST 作为全量传输,但也只是使用 XtraBackup 方式。

节点在集群中,会因为新节点的加入或故障,同步失效等而发生状态的切换,下面列举出这状态的含义。

① open:节点启动成功,尝试连接到集群。

② primary:节点已在集群中,在新节点加入集群时,选取 doner 进行数据同步时会产生式的状态。

③ joiner:节点处于等待接收同步数据文件的状态。

④ joined:节点已完成了数据同步,尝试保持和集群中其他节点进度一致。

⑤ synced:节点正常提供服务的状态,表示已经同步完成并和集群进度保持一致。

⑥ doner:节点处于为新加入节点提供全量数据时的状态。

16.4.2 参数配置

1. 搭建 PXC 过程中,需要在 my.cnf 中设置以下参数

wsrep_cluster_name:指定集群的逻辑名称,对于集群中的所有节点,名称必须相同。

wsrep_cluster_address:指定集群中各节点地址。

wsrep_node_name:指定当前节点在集群中的逻辑名称。

wsrep_node_address:指定当前节点的 IP。

wsrep_provider:指定 galera 库的路径。

wsrep_sst_method:模式情况下,PXC 使用 XtraBackup 进行 SST 传输。强烈建议 failure 参数指为 xtrabackup-v2。

wsrep_sst_auth:指定认证凭证 SST 作为 sst_user:sst_pwd。必须在引导第一个节点后创建此用户并赋予必要的权限。PXC8.0 版本中 wsrep_sst_auth 这个参数已经被移除了,不用设置。

pxc_strict_mode:严格模式,官方建议该参数值为 enforcing。

2. gcache

在 PXC 中还有一个特别重要的模块就是 gcache。它的核心功能就是每个节点缓存当前最新的写集。如果有新节点加入集群,就可以把新数据等待增量传输给新节点,而不需要使用 SST 方式。这样可以让节点更快加入到集群中。

gcache 模块涉及了如下参数:

gcache.size:代表用来缓存写集增量信息的大小。它的默认大小为 128 MB,通过 wsrep_provider_options 变量参数设置。建议调整为 2G-4G 范围,足够的空间便于缓存更多的增量信息。

gcache.mem.size:代表 gcache 中内存缓存的大小,适度调大可以提高整个集群性能。

gcache.page.size:可以理解为如果内存不够用(cache 不足),就直接将写集写入到磁盘文件中。

16.4.3 PXC 集群状态监控

在集群搭建好之后,可以通过以下状态变量'% wsrep %'来查看集群各节点的状态。

wsrep_local_state_uuid:集群中所有节点的该状态值应该是相同的,如果有不同值的节点,说明其没有加入集群。

wsrep_last_committed:最后提交的事务数目。

wsrep_cluster_size:当前集群中的节点数量。

wsrep_cluster_status:集群组成的状态。如果不是"primary",说明出现脑裂现象。

wsrep_local_state:当前节点状态,值为 4 表示正常。该状态有 4 个值。

joining:表示节点正在加入集群。

doner:节点处于为新加入节点提供全量数据时的状态。

joined:当前节点已成功加入集群。

synced:当前节点与集群中各节点是同步状态。

wsrep_ready:为 on 表示当前节点可以正常提供服务。为 off 表示节点可能发生脑裂或网络问题导致。

16.4.4 grastate.dat

grastate.dat 是节点状态记录文件,它一个非常重要的文件。节点故障恢复的第一步就是恢复这个文件,节点重新加入集群第一个需要关注的也是这个文件。这个文件位于数据目录下。该文件的内容如下:

uuid:节点所属集群的 uuid。

seqno:节点正常关闭时最后提交的事务编号,如果集群非正常关闭或者正在运行状态,seqno 的值为-1。

safe_to_bootstrap:集群最后一个关闭的节点的值为 1,其他为 0。

节点重新加入集群时,会从 grastat.dat 中的 seqno 开始做 IST。若这个文件不存在或者损坏了,那么意味着节点将会以 SST 的方式加入集群,这是我们最不愿见到的。

16.5 PXC 高可用集群部署

16.5.1 架构规划

本次部署基于 Centos8 的操作系统部署 PXC 8.0 的版本,架构规划如图 16-2 所示,如表 16-1 列。

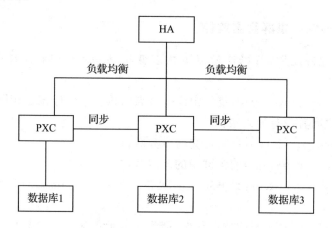

图 16-2 PXC 高可用集群

表 16-1 PXC 服务器架构规划

角色	ip 地址	主机名	server_id	类型
PXC	192.168.0.112	pxc-node1	803241	PXC 节点
PXC	192.168.0.113	pxc-node2	803242	PXC 节点
PXC	192.168.0.114	pxc-node3	803243	PXC 节点
HA	192.168.0.115	haproxy	—	负载均衡

部署中用到安装包:

网址:https://www.percona.com/downloads/

PXC:Percona-XtraDB-Cluster_8.0.28-19.1_Linux.x86_64.glibc2.17.tar.gz

PXB:percona-xtrabackup-8.0.28-20-Linux-x86_64.glibc2.17.tar.gz

网址:http://repo.percona.com/pxc-80/yum/release/

压缩备份与还原工具 Qpress:qpress-11-3.el8.x86_64.rpm

16.5.2 本地 yum 源配置

[root@pxc-node1 ~]# cat /etc/redhat-release
CentOS Linux release 8.2.2004 (Core)

(1) yum 配置中

注意：注销 mirrorlist，添加以下内容，3 个节点都做如下操作。

[root@pxc-node1 ~]# cd /etc/yum.repos.d
[root@pxc-node1 ~]# vi CentOS-Base.repo
baseurl = https://vault.centos.org/centos/$releasever/BaseOS/$basearch/os/
[root@pxc-node1 ~]# vi CentOS-AppStream.repo
baseurl = https://vault.centos.org/centos/$releasever/AppStream/$basearch/os/

(2) 安装依赖包

yum -y install perl perl-devel libaio libaio-devel perl-Time-HiRes perl-DBD-MySQL numactl
yum install net-tools
yum install -y libtinfo*
yum install -y libev
yum install -y openssl socat procps-ng chkconfig shadow-utils **coreutils**
yum install diff*

16.5.3 PXB 安装

1. 解压缩安装包，三个节点均安装

-- **pxc-node1**
tar -zxvf percona-xtrabackup-8.0.28-20-Linux-x86_64.glibc2.17.tar.gz
mv percona-xtrabackup-8.0.28-20-Linux-x86_64.glibc2.17 /usr/local/percona-xtrabackup

2. 常见软链接

ln -s /usr/local/percona-xtrabackup/bin/xtrabackup /usr/bin/xtrabackup
ln -s /usr/local/percona-xtrabackup/bin/xbstream /usr/bin/xbstream
[root@pxc-node1 /]# which xtrabackup
/usr/bin/xtrabackup

―― pxc-node2

```
tar -zxvf percona-xtrabackup-8.0.28-20-Linux-x86_64.glibc2.17.tar.gz
mv percona-xtrabackup-8.0.28-20-Linux-x86_64.glibc2.17 /usr/local/percona-xtrabackup
```

3. 常见软链接

```
ln -s /usr/local/percona-xtrabackup/bin/xtrabackup /usr/bin/xtrabackup
ln -s /usr/local/percona-xtrabackup/bin/xbstream /usr/bin/xbstream
[root@pxc-node2 /]# which xtrabackup
/usr/bin/xtrabackup
```

―― pxc-node3

```
tar -zxvf percona-xtrabackup-8.0.28-20-Linux-x86_64.glibc2.17.tar.gz
mv percona-xtrabackup-8.0.28-20-Linux-x86_64.glibc2.17 /usr/local/percona-xtrabackup
```

4. 常见软链接

```
ln -s /usr/local/percona-xtrabackup/bin/xtrabackup /usr/bin/xtrabackup
ln -s /usr/local/percona-xtrabackup/bin/xbstream /usr/bin/xbstream
[root@pxc-node3 /]# which xtrabackup
/usr/bin/xtrabackup
```

注意：xtrabackup --version 可以查询 PXB 的版本。

16.5.4　PXC 安装

1. qpress 工具部署

三个节点均做如下操作：

```
[root@pxc-node1 /]# rpm -ivh qpress-11-3.el8.x86_64.rpm
[root@pxc-node1 /]# which qpress
/usr/bin/qpress
```

2. 用户组及用户添加

三个节点均做如下操作，解压后目录如图 16-3 所示。

```
groupadd mysql
useradd -r -g mysql mysql
tar -zxvf Percona-XtraDB-Cluster_8.0.28-19.1_Linux.x86_64.glibc2.17.tar.gz
mv Percona-XtraDB-Cluster_8.0.28-19.1_Linux.x86_64.glibc2.17 /usr/
```

local/pxc-8028

3. 环境变量设置,三个节点均做如下操作

[root@pxc-node1 /]# chown -R mysql.mysql /usr/local/pxc-8028

[root@pxc-node1 /]# echo "export PATH=/usr/local/pxc-8028/bin:$PATH" >> /root/.bashrc

[root@pxc-node1 /]# source /root/.bashrc

```
[root@pxc-node1 /]# ls -l /usr/local/pxc-8028
total 652
-rw-r--r--.   1 root root   1703 Jul 13 09:04 COPYING-jemalloc
-rw-r--r--.   1 root root 276595 Jul 13 07:39 LICENSE
-rw-r--r--.   1 root root 276595 Jul 13 07:39 LICENSE-test
-rw-r--r--.   1 root root  47676 Jul 13 07:39 LICENSE.router
-rw-r--r--.   1 root root  19626 Jul 13 07:39 README-wsrep
-rw-r--r--.   1 root root   2153 Jul 13 07:39 README.md
-rw-r--r--.   1 root root   2153 Jul 13 07:39 README.md-test
-rw-r--r--.   1 root root    679 Jul 13 07:39 README.router
drwxr-xr-x.   3 root root   4096 Jul 13 09:04 bin
drwxr-xr-x.   2 root root     69 Jul 13 09:04 cmake
drwxr-xr-x.   2 root root     69 Jul 13 09:04 docs
drwxr-xr-x.   6 root root   4096 Jul 13 09:04 include
drwxr-xr-x.   7 root root   4096 Jul 13 09:04 lib
drwxr-xr-x.   4 root root     30 Jul 13 09:04 man
drwxr-xr-x.  10 root root   4096 Jul 13 09:04 mysql-test
-rw-r--r--.   1 root root   1617 Jul 13 08:31 mysqlrouter-log-rotate
drwxr-xr-x.   5 root root     50 Jul 13 09:04 percona-xtradb-cluster-tests
drwxrwxrwx.   2 root root      6 Jul 13 09:04 run
drwxr-xr-x.  28 root root   4096 Jul 13 09:04 share
drwxr-xr-x.   2 root root     77 Jul 13 09:04 support-files
drwxr-xr-x.   3 root root     17 Jul 13 09:04 var
```

图 16-3 PXC 安装后目录

4. 参数文件设置

--pxc-node1

[root@pxc-node1 /]# cat > /etc/my.cnf << EOF

[mysqld]

basedir=/usr/local/pxc-8028

datadir=/usr/local/pxc-8028/data

user=mysql

port=3306

server_id=803241

binlog_format=ROW

character_set_server=utf8mb4

skip-name-resolve

wsrep_provider=/usr/local/pxc-8028/lib/libgalera_smm.so

wsrep_cluster_address=gcomm://192.168.0.112,192.168.0.113,192.168.0.114

wsrep_sst_method=xtrabackup-v2

wsrep_cluster_name=jem_pxc_cluster

```
wsrep_slave_threads = 4
wsrep_node_name = pxc-node1
wsrep_node_address = 192.168.0.112
EOF
```

-- pxc-node2

```
[root@pxc-node2 /]# cat > /etc/my.cnf << EOF
[mysqld]
basedir = /usr/local/pxc-8028
datadir = /usr/local/pxc-8028/data
user = mysql
port = 3306
server_id = 803242
binlog_format = ROW
character_set_server = utf8mb4
skip-name-resolve
wsrep_provider = /usr/local/pxc-8028/lib/libgalera_smm.so
wsrep_cluster_address = gcomm://192.168.0.112,192.168.0.113,192.168.0.114
wsrep_sst_method = xtrabackup-v2
wsrep_cluster_name = jem_pxc_cluster
wsrep_slave_threads = 4
wsrep_node_name = pxc-node2
wsrep_node_address = 192.168.0.113
EOF
```

-- pxc-node3

```
[root@pxc-node3 /]# cat > /etc/my.cnf << EOF
[mysqld]
basedir = /usr/local/pxc-8028
datadir = /usr/local/pxc-8028/data
user = mysql
port = 3306
server_id = 803243
binlog_format = ROW
character_set_server = utf8mb4
skip-name-resolve
wsrep_provider = /usr/local/pxc-8028/lib/libgalera_smm.so
```

wsrep_cluster_address = gcomm://192.168.0.112,192.168.0.113,192.168.0.114
wsrep_sst_method = xtrabackup-v2
wsrep_cluster_name = jem_pxc_cluster
wsrep_slave_threads = 4
wsrep_node_name = pxc-node3
wsrep_node_address = 192.168.0.114
EOF

重要参数说明：

(1)wsrep_sst_auth：与 Percona XtraDB Cluster 5.7 不同，PXC8.0 的版本中 wsrep_sst_auth 这个参数已经被移除了，不用设置。

(2)wsrep_provider：指定 Galera 库的路径。

(3)wsrep_cluster_address：指定群集中节点的 IP 地址。

(4)wsrep_sst_method：Percona XtraDB 群集使用 Percona XtraBackup 进行状态快照传输。

(5)wsrep_cluster_name：wsrep_集群名称，所有节点保持一致。

(6)binlog_format：Galera 只支持行级复制，因此设置 binlog_format 为 row。

16.5.5 PXC 初始化

1. 只初始化 pxc-node1 数据库

pxc-node1 数据库成功初始化如图 16-4 所示。

[root@pxc-node1 /]# /usr/local/pxc-8028/bin/mysqld --initialize-insecure --user=mysql \
--basedir=/usr/local/pxc-8028 --datadir=/usr/local/pxc-8028/data

```
[root@pxc-node1 /]# /usr/local/pxc-8028/bin/mysqld --initialize-insecure --user=mysql --basedir=/usr/local/pxc-8028 --datadir=/usr/local/pxc-8028/data
2022-12-11T11:04:31.673285Z 0 [Warning] [MY-011068] [Server] The syntax 'wsrep_slave_threads' is deprecated and will be removed in a future release. Please use wsrep_applier_threads instead.
2022-12-11T11:04:31.673352Z 0 [Warning] [MY-000000] [WSREP] Node is running in bootstrap/initialize mode. Disabling pxc_strict_mode checks
2022-12-11T11:04:31.674058Z 0 [System] [MY-013169] [Server] /usr/local/pxc-8028/bin/mysqld (mysqld 8.0.28-19.1) initializing of server in progress as process 424
2022-12-11T11:04:31.766168Z 0 [Note] [MY-000000] [Galera] Loading provider none initial position: 00000000-0000-0000-0000-000000000000:-1
2022-12-11T11:04:31.766246Z 0 [Note] [MY-000000] [Galera] wsrep_load(): loading provider library 'none'
2022-12-11T11:04:31.839961Z 1 [System] [MY-013576] [InnoDB] InnoDB initialization has started.
2022-12-11T11:04:33.078246Z 1 [System] [MY-013577] [InnoDB] InnoDB initialization has ended.
2022-12-11T11:04:33.948226Z 6 [Warning] [MY-010453] [Server] root@localhost is created with an empty password ! Please consider switching off the --initialize-insecure option.
[root@pxc-node1 /]#
```

图 16-4 PXC 初始化成功

--启动 pxc-node1,初始化成功如图 16-5 所示。

mysqld_safe --defaults-file=/etc/my.cnf --ledir=/usr/local/pxc-8028/bin/ --wsrep-new-cluster &

--告警日志文件查看,启动成功的日志如图 16-6 所示。

```
[root@pxc-node1 /]# mysqld_safe --defaults-file=/etc/my.cnf --ledir=/usr/local/pxc-8028/bin/ --wsrep-new-cluster &
[1] 472
[root@pxc-node1 /]# 2022-12-11T11:05:20.212940Z mysqld_safe Logging to '/usr/local/pxc-8028/data/pxc-node1.err'.
Logging to '/usr/local/pxc-8028/data/pxc-node1.err'.
2022-12-11T11:05:20.271367Z mysqld_safe Starting mysqld daemon with databases from /usr/local/pxc-8028/data
2022-12-11T11:05:20.288167Z mysqld_safe Skipping wsrep-recover for empty datadir: /usr/local/pxc-8028/data
2022-12-11T11:05:20.289654Z mysqld_safe Assigning 00000000-0000-0000-0000-000000000000:-1 to wsrep_start_position
```

图 16-5 PXC 集群启动成功

[root@pxc-node1 /]# tail-f /usr/local/pxc-8028/data/pxc-node1.err

```
[root@pxc-node1 /]# tail -f /usr/local/pxc-8028/data/pxc-node1.err
2022-12-11T11:05:24.612984Z 2 [Note] [MY-000000] [WSREP] Server status change initialized -> joined
2022-12-11T11:05:24.612990Z 2 [Note] [MY-000000] [WSREP] wsrep_notify_cmd is not defined, skipping notification.
2022-12-11T11:05:24.613003Z 2 [Note] [MY-000000] [WSREP] wsrep_notify_cmd is not defined, skipping notification.
2022-12-11T11:05:24.618155Z 2 [Note] [MY-000000] [Galera] Recording CC from group: 1
2022-12-11T11:05:24.618174Z 2 [Note] [MY-000000] [Galera] Lowest cert index boundary for CC from group: 1
2022-12-11T11:05:24.618180Z 2 [Note] [MY-000000] [Galera] Min available from gcache for CC from group: 1
2022-12-11T11:05:24.618233Z 2 [Note] [MY-000000] [Galera] Server pxc12 synced with group
2022-12-11T11:05:24.618243Z 2 [Note] [MY-000000] [WSREP] Server status change joined -> synced
2022-12-11T11:05:24.618247Z 2 [Note] [MY-000000] [WSREP] Synchronized with group, ready for connections
2022-12-11T11:05:24.618250Z 2 [Note] [MY-000000] [WSREP] wsrep_notify_cmd is not defined, skipping notification.
```

图 16-6 PXC 集群数据库启动成功日志

——登录 MySQL，登录成功后，如图 16-7 所示。

```
[root@pxc-node1 /]# mysql -S/tmp/mysql.sock
Welcome to the MySQL monitor.  Commands end with ; or \g.
Your MySQL connection id is 14
Server version: 8.0.28-19.1 Percona XtraDB Cluster binary (GPL) 8.0.28, Revision f544540, WSREP version 26.4.3

Copyright (c) 2009-2022 Percona LLC and/or its affiliates
Copyright (c) 2000, 2022, Oracle and/or its affiliates.

Oracle is a registered trademark of Oracle Corporation and/or its
affiliates. Other names may be trademarks of their respective
owners.

Type 'help;' or '\h' for help. Type '\c' to clear the current input statement.

mysql> select host,user from mysql.user;
+-----------+---------------------------+
| host      | user                      |
+-----------+---------------------------+
| localhost | mysql.infoschema          |
| localhost | mysql.pxc.internal.session|
| localhost | mysql.pxc.sst.role        |
| localhost | mysql.session             |
| localhost | mysql.sys                 |
| localhost | root                      |
+-----------+---------------------------+
6 rows in set (0.00 sec)

mysql> status
--------------
mysql  Ver 8.0.28-19.1 for Linux on x86_64 (Percona XtraDB Cluster binary (GPL) 8.0.28, Revision f544540, WSREP version 26.4.3)

Connection id:          14
Current database:
Current user:           root@localhost
SSL:                    Not in use
Current pager:          stdout
Using outfile:          ''
Using delimiter:        ;
Server version:         8.0.28-19.1 Percona XtraDB Cluster binary (GPL) 8.0.28, Revision f544540, WSREP version 26.4.3
Protocol version:       10
Connection:             Localhost via UNIX socket
Server characterset:    utf8mb4
Db     characterset:    utf8mb4
Client characterset:    utf8mb4
Conn.  characterset:    utf8mb4
UNIX socket:            /tmp/mysql.sock
```

图 16-7 PXC 集群启动成功

[root@pxc-node1 /]# mysql -S/tmp/mysql.sock

mysql> select host,user from mysql.user

——端口监听

[root@pxc-node1 /]# netstat -lntup|egrep "4567|3306"

——修改 root 本地密码

mysql>alter user root@'localhost' identified with mysql_native_password by 'root'

```
mysql> flush privileges
```

--创建远程用户root密码

```
mysql> create user root@'%' identified with mysql_native_password by 'root'
mysql> grant all on *.* to root@'%' with grant option
mysql> flush privileges
mysql> show status like 'wsrep_cluster%'
```

Variable_name	Value
wsrep_cluster_weight	1
wsrep_cluster_capabilities	
wsrep_cluster_conf_id	1
wsrep_cluster_size	1
wsrep_cluster_state_uuid	b446c11f-7943-11ed-9501-a6234abbb34b
wsrep_cluster_status	Primary

16.5.6　PXC添加节点

```
## PXC集群添加节点pxc-node2
[root@pxc-node2 /]# mkdir -p /usr/local/pxc-8028/data
[root@pxc-node2 /]# chown -R mysql:mysql /usr/local/pxc-8028/data
[root@pxc-node2 /]# mysqld_safe --defaults-file=/etc/my.cnf --ledir=/usr/local/pxc-8028/bin/ &
```

--默认密钥和证书文件复制到第二个节点

```
[root@pxc-node1 ~]# cd /usr/local/pxc-8028/data
[root@pxc-node1 data]# scp *.pem pxc-node2:/usr/local/pxc-8028/data
## PXC集群添加节点pxc-node3
[root@pxc-node3 /]# mkdir -p /usr/local/pxc-8028/data
[root@pxc-node3 /]# chown -R mysql:mysql /usr/local/pxc-8028/data
[root@pxc-node3 /]# mysqld_safe --defaults-file=/etc/my.cnf --ledir=/usr/local/pxc-8028/bin/ &
```

--默认密钥和证书文件复制到第三个节点

```
[root@pxc-node1 ~]# cd /usr/local/pxc-8028/data
```

```
[root@pxc-node1 data]# scp *.pem pxc-node3:/usr/local/pxc-8028/data
```

根据官网资料，pxc-encrypt-cluster-traffic 变量默认是启用，所以要求所有节点使用相同的密钥和证书文件，MySQL 生成默认密钥和证书文件，并将其放在数据目录中。这些自动生成的文件适用于自动 SSL 配置。所以将第一个节点数据目录下的 *.pem 全部复制到同步的节点。目的也是有助于提高系统的安全性。如果禁用 pxc-encrypt 集群秘钥传输，则需要停止集群并更新配置文件的[mysqld]部分：每个节点的 pxc-encrypt-cluster-traffic=OFF。然后，重新启动群集即可。

此时查询 PXC 集群为 3 个节点，多主模式，信息如下：

```
mysql> show status like 'wsrep_cluster_%';
+--------------------------+--------------------------------------+
| Variable_name            | Value                                |
+--------------------------+--------------------------------------+
| wsrep_cluster_weight     | 3                                    |
| wsrep_cluster_capabilities |                                    |
| wsrep_cluster_conf_id    | 7                                    |
| wsrep_cluster_size       | 3                                    |
| wsrep_cluster_state_uuid | ec040eed-7965-11ed-98d1-ffd3143d8ea6 |
| wsrep_cluster_status     | Primary                              |
+--------------------------+--------------------------------------+
```

16.5.7 校验 Replication

```
mysql@pxc-node1> CREATE DATABASE percona
mysql@pxc-node1> USE percona
mysql@pxc-node1> CREATE TABLE example
(node_id INT PRIMARY KEY, node_name VARCHAR(30))
mysql@pxc-node1> INSERT INTO percona.example VALUES (1, 'percona1')

##此时在第二个和第三个节点查询，信息同步
mysql@pxc-node3> SELECT * FROM percona.example
+---------+-----------+
| node_id | node_name |
+---------+-----------+
|       1 | percona1  |
+---------+-----------+
```

16.5.8 关闭重启

如果三台都关闭了,那么要先启动最后关闭的那台机器的服务,第一个启动的节点需要加上――wsrep-new-cluster。如果忘记了是哪台最后关闭,此时通过grastate.dat这个文件就可以识别出来,集群最后一个关闭的节点的值为1,其他为0。

[root@pxc-node1 ~]# more /usr/local/pxc-8028/data/grastate.dat
GALERA saved state
version: 2.1
uuid: ec040eed-7965-11ed-98d1-ffd3143d8ea6
seqno: -1
safe_to_bootstrap: 0
[root@pxc-node2 ~]# more /usr/local/pxc-8028/data/grastate.dat
GALERA saved state
version: 2.1
uuid: ec040eed-7965-11ed-98d1-ffd3143d8ea6
seqno: -1
safe_to_bootstrap: 0
[root@pxc-node3 ~]# more /usr/local/pxc-8028/data/grastate.dat
GALERA saved state
version: 2.1
uuid: ec040eed-7965-11ed-98d1-ffd3143d8ea6
seqno: -1
safe_to_bootstrap:1

通过grastate.dat文件记录的信息,第三个节点为最后关闭的。

16.6 HAProxy

16.6.1 HAProxy介绍

HAProxy是一款提供高可用性、负载均衡以及基于TCP(第4层)和HTTP(第7层)应用的代理软件,支持虚拟主机,它是免费、快速并且可靠的一种解决方案。HAProxy特别适用于那些负载特大的web站点,这些站点通常又需要会话保持或7层处理。HAProxy运行在时下的硬件上,完全可以支持数以万计的并发连接,并且它的运行模式使得它可以很简单安全整合进您当前的架构中,同时可以保护你的web服务器不被暴露到网络上。

官方网址：https://www.haproxy.com/，官网主页如图 16-8 所示

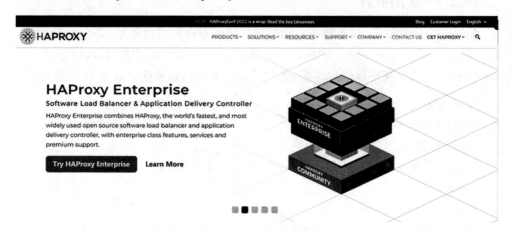

图 16-8　HAProxy 官网主页

16.6.2　yum 安装 HAProxy

1. 获取最新的 haproxy 源

[root@haproxy ~]# yum install centos-release-scl -y

2. yum 在线安装

[root@haproxy ~]# yum install -y haproxy

16.6.3　HAProxy 配置文件

cat > /etc/haproxy/haproxy.cfg <<"EOF"
global
log /dev/log local0
log /dev/log local1 notice
chroot /var/lib/haproxy
user haproxy
group haproxy
daemon　#以后台形式运行 ha-proxy
defaults
log global
mode tcp
option dontlognull
timeout connect 5000

```
timeout client 1m
timeout server 1m
timeout http-keep-alive 10s
timeout check 10s
```

1. web 页面访问

```
listen status
mode http
bind *:33060
stats enable
stats uri /admin
stats auth admin:admin  #用户认证
stats admin if TRUE
stats realm Haproxy\statistics
```

2. 访问 MySQL 的配置

```
frontend main
bind 0.0.0.0:3306  #监听哪个 ip 和什么端口
default_backend mysql  #默认使用的服务器组
backend mysql
balance roundrobin   #负载均衡的方式
server PXC1 192.168.0.112:3306 check inter 2000 rise 3 fall 3 weight 30
server PXC2 192.168.0.113:3306 check inter 2000 rise 3 fall 3 weight 30
server PXC3 192.168.0.114:3306 check inter 2000 rise 3 fall 3 weight 30
EOF
```

16.6.4　HAProxy 管理

1. 配置文件校验

```
[root@haproxy ~]# haproxy -f /etc/haproxy/haproxy.cfg -c
Configuration file is valid
```

2. 进程管理

```
[root@haproxy ~]# systemctl enable haproxy.service
Created symlink /etc/systemd/system/multi-user.target.wants/haproxy.service → /usr/lib/systemd/system/haproxy.service.
[root@haproxy ~]# systemctl restart haproxy.service
```

```
[root@haproxy ~]# systemctl status haproxy.service
  haproxy.service - HAProxy Load Balancer
  Loaded: loaded (/usr/lib/systemd/system/haproxy.service; enabled;
vendor preset: disabled)
  Active: active (running) since Mon 2022-12-12 15:30:22 UTC; 6s ago
  Process: 225 ExecStartPre=/usr/sbin/haproxy -f $CONFIG -c -q $OPTIONS
(code=exited, status=0/SUCCESS)
  Main PID: 227 (haproxy)
    Tasks: 2 (limit: 24036)
   Memory: 1.7M
   CGroup: /docker/aef65e105ab762d838eb9d659707e92f3303cf88de22436de333531e238569af/system.slice/haproxy.service
           ├─227 /usr/sbin/haproxy -Ws -f /etc/haproxy/haproxy.cfg -p /run/haproxy.pid
           └─229 /usr/sbin/haproxy -Ws -f /etc/haproxy/haproxy.cfg -p /run/haproxy.pid

Dec 12 15:30:22 haproxy systemd[1]: Starting HAProxy Load Balancer...
Dec 12 15:30:22 haproxy haproxy[227]: Proxy status started.
Dec 12 15:30:22 haproxy haproxy[227]: Proxy status started.
Dec 12 15:30:22 haproxy haproxy[227]: Proxy main started.
Dec 12 15:30:22 haproxy haproxy[227]: Proxy main started.
Dec 12 15:30:22 haproxy haproxy[227]: Proxy mysql started.
Dec 12 15:30:22 haproxy haproxy[227]: Proxy mysql started.
Dec 12 15:30:22 haproxy systemd[1]: Started HAProxy Load Balancer.
[root@haproxy ~]# netstat -plantu | grep 3306
tcp    0    0 0.0.0.0:3306    0.0.0.0:*    LISTEN    229/haproxy
tcp    0    0 0.0.0.0:33060   0.0.0.0:*    LISTEN    229/haproxy
```

3. web 访问

http://ip:33060/admin,网页监控如图 16-9 所示。

http://192.168.1.54:33060/admin

4. 读负载均衡测试

如图 16-10 所示,根据配置文件的权重连接。

```
[root@pxc-node2 ~]# for i in $(seq 1 10); do mysql -uroot -proot \
-h192.168.0.115 -P3306 -e 'select @@server_id;'; done | egrep '[0-9]'
```

第 16 章 高可用之 PXC 架构

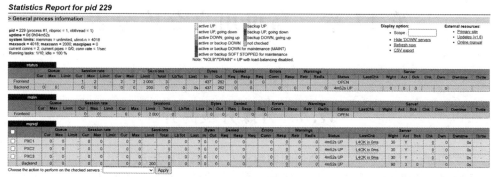

图 16-9 HAProxy 监控界面

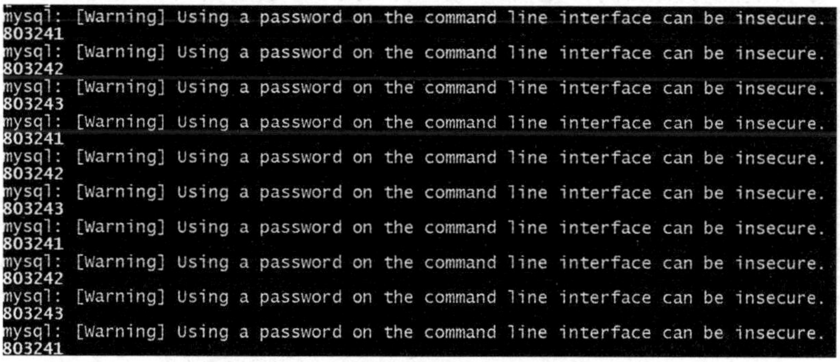

图 16-10 连接 HAProxy 负载均衡测试

第 17 章 高可用之分库分表

17.1 分库分表简介

MySQL 作为互联网公司都会用到的数据库,如果在使用过程中出现性能问题,会采用 MySQL 的横向扩展,使用主从复制来提高读性能,要是解决写入问题,需要进行分库分表。

分库分表是业务发展到一定阶段,数据积累到一定量级而衍生出来的解决方案。当 DB 的数据量级到达一个阶段,写入和读取的速度会出现瓶颈,即使是有索引,索引也会变很大,而且数据库的物理文件大的会使备份和恢复等操作变得很困难。这个时候由于 DB 的瓶颈已经严重危害到了业务,最有效的解决方案莫过于 DB 的分库分表了。

数据库表拆分解决的问题主要是存储和性能问题,MySQL 在单表数据量达到一定量级后,性能会急剧下降,相比于 sqlserver 和 Oracle 这些收费 DB 来说,MySQL 在某些方面还是处于弱势,但是表的拆分这个策略却适用于几乎所有的关系型数据库。

17.2 分库分表的目的

一个系统,目前订单数据量已达上亿,并且每日以百万级别的速度增长,甚至之后还可能是千万级。面对如此庞大的数据量,那么一旦数据量疯狂增长,必然造成读写缓慢。那么,为了使系统能够抗住千万级数据量的压力,都有哪些解决方案呢?分库分表为首选。

分库分表就是为了解决由于数据量过大而导致数据库性能降低的问题,将原来独立的数据库拆分成若干数据库组成,将数据大表拆分成若干数据表组成,使得单一数据库、单一数据表的数据量变小,从而达到提升数据库性能的目的,就需要用到一些数据库中间件。

17.3 常用数据库中间件

17.3.1 Mycat

MyCAT 是一个彻底开源的,面向企业应用开发的"大数据库集群"支持事务、ACID、可以替代 Mysql 的加强版数据库?一个可以视为"Mysql"集群的企业级数据库,用来替代昂贵的 Oracle 集群?一个融合内存缓存技术、Nosql 技术、HDFS 大数据的新型 SQL Server?结合传统数据库和新型分布式数据仓库的新一代企业级数据库产品?一个新颖的数据库中间件产品不仅可以用作读写分离,以及分表分库、容灾管理,而且可以用于多租户应用开发、云平台基础设施,让你的架构具备很强的适应性和灵活性。

MyCAT 里面通过定义路由规则来实现分片表(路由规则里面会定义分片字段,以及分片算法)。分片算法有多种,hash 是其中一种,还有取模、按范围分片等等。在 mycat 里面,会对所有传递的 sql 语句做路由处理(路由处理的依据就是表是否分片,如果分片,那么需要依据分片字段和对应的分片算法来判断 sql 应该传递到哪一个,哪几个以及全部节点去执行。

17.3.2 Cobar

Cobar 是提供关系型数据库(MySQL)分布式服务的中间件,它可以让传统的数据库得到良好的线性扩展,并看上去还是一个数据库,对应用保持透明。Cobar 属于中间层方案,在应用程序和 MySQL 之间搭建一层 Proxy。中间层介于应用程序与数据库间,需要做一次转发,而基于 JDBC 协议并无额外转发,直接由应用程序连接数据库,性能上有些许优势。这里并非说明中间层一定不如客户端直连,除了性能,需要考虑的因素还有很多,中间层更便于实现监控、数据迁移、连接管理等功能。

17.4 分库分表的类型

(1)分库:垂直分库、水平分库。
(2)分表:垂直分表、水平分表。
分库是指把一个数据库拆分为多个数据库,一般分为垂直分库和水平分库。
分表指的是通过一定规则,将一张表分解成多张不同的表,一般分为垂直分表和水平分表。

17.4.1 垂直分库

以表为依据,按照业务归属不同,将不同的表拆分到不同的库中,每个库可以放在不同的服务器上,核心理念是专库专用。随着业务的发展一些公用的配置表、字典表等越来越多,这时可以将这些表拆到单独的库中,甚至可以服务化。随着业务的发展孵化出了一套业务模式,这时可以将相关的表拆到单独的库中,甚至可以服务化。

垂直分库的结果是:
(1)每个库的表结构都不一样。
(2)每个库的数据也不一样,没有交集。
(3)所有库的并集是全量数据。

17.4.2 水平分库

以字段为依据,按照一定策略(hash、range 等),将一个库中的数据拆分到多个库中。这样库多了,IO 和 CPU 的压力自然可以成倍缓解。

水平分库的结果是:
(1)每个库的结构都一样。
(2)每个库的数据都不一样,没有交集。
(3)所有库的并集是全量数据。

17.4.3 垂直分表

垂直分表即"宽表拆窄表",以字段为依据,按照字段的活跃性,将表中字段拆到不同的表(主表和扩展表)中。

系统绝对并发量并没有上来,表的记录并不多,但是字段多,并且热点数据和非热点数据在一起,单行数据所需的存储空间较大。以至于数据库缓存的数据行减少,查询时会去读磁盘数据产生大量的随机读 IO,产生 IO 瓶颈,这时候垂直分表就可以解决这个问题。拆了之后,要想获得全部数据就需要关联两个表来取数据。

17.4.4 水平分表

水平分表是以字段为依据,按照一定策略(hash、range 等),将一个表中的数据拆分到多个表中,也称为库内分表,水平分表的结果是:
(1)每个表的结构都一样。
(2)每个表的数据都不一样,没有交集。
(3)所有表的并集是全量数据。

17.5 小 结

分库分表的顺序应该是先垂直分,后水平分,先垂直分表,再垂直分库,再水平分库,最后水平分表。因为垂直分更简单,更符合业务场景。

第 17 章　高可用之分库分表

有 DBA 的朋友，常问起，分库分表和表分区有什么区别？其实它们的本质区别是：

（1）表分区（Partitioning）可以将一张表的数据分别存储为多个文件，从逻辑上来讲只有一张表，是 MySQL 的一种内部实现。

（2）分表则是将一张表分解成多张表，分库分表需要代码实现。分库分表和分区并不冲突，可以结合使用。

附录1 8.0.30 or Higher 新特性

1. Redo Log

1.1 innodb_redo_log_capacity 参数

在 MySQL 8.0.30 中，innodb_redo_log_capacity 系统变量控制重做日志文件占用的磁盘空间量。可以在启动或运行时使用 set GLOBAL 语句在选项文件中设置此变量；例如，以下语句将重做日志容量设置为 8 GB：

SET GLOBAL innodb_redo_log_capacity = 8589934592;

说明：

(1)innodb_redo_log_capacity 变量取代了已弃用的
innodb_log_files_in_group 和 innodb_log_file_size 变量。

定义 innodb_redo_log_capacity 设置时，将忽略 innodb_log_files_in_group 和 innodb_log_file_size 设置，否则，这些设置将用于计算 innodb_redo_log_capacity 设置，公式为：

innodb.log_files_in_group * innodblog_file_size = innodb_do_log_capacity

(2)如果没有设置这些变量，则重做日志容量将设置为 innodb_redo_log_capacity 默认值，即 104 857 600 字节(100 MB)，最大重做日志容量为 128 GB。

1.2 重做日志文件

在 MySQL 8.0.30 之前，InnoDB 默认在数据目录中创建两个重做日志文件，分别名为 ib_logfile0 和 ib_logfile1，并以循环方式写入这些文件。

重做日志文件使用 #ib_redoN 命名约定，其中 N 是重做日志的文件号。备用重做日志文件由_tmp 后缀表示。下面的附录图 1 显示了 #innodb_redo 目录中的重做日志文件，其中有 1 个活动重做日志和 31 个备用重做日志，按顺序编号。

除非 innodb_log_group_home_dir 变量指定了不同的目录，否则重做日志文件位于数据目录的 #innodb_Redo 目录中。如果定义了 innodb_log_group_home_dir，则重做日志文件位于该目录中的 #innodb_redo 目录中。

有两种类型的重做日志文件，普通和备用。普通的重做日志文件就是正在使用的那些文件。备用重做日志文件是那些等待使用的文件。

InnoDB 尝试维护总共 32 个重做日志文件,每个文件的大小等于 1/32 * InnoDB_redo_log_capacity

```
mysql> select @@innodb_log_group_home_dir
+-----------------------------+
| @@innodb_log_group_home_dir |
+-----------------------------+
| ./                          |
+-----------------------------+
1 row in set (0.00 sec)
```

```
total 102400
-rw-r----- 1 mysql mysql 3276800 Nov 11 13:12 '#ib_redo10_tmp'
-rw-r----- 1 mysql mysql 3276800 Nov 11 13:12 '#ib_redo11_tmp'
-rw-r----- 1 mysql mysql 3276800 Nov 11 13:12 '#ib_redo12_tmp'
-rw-r----- 1 mysql mysql 3276800 Nov 11 13:12 '#ib_redo13_tmp'
-rw-r----- 1 mysql mysql 3276800 Nov 11 13:12 '#ib_redo14_tmp'
-rw-r----- 1 mysql mysql 3276800 Nov 11 13:12 '#ib_redo15_tmp'
-rw-r----- 1 mysql mysql 3276800 Nov 11 13:12 '#ib_redo16_tmp'
-rw-r----- 1 mysql mysql 3276800 Nov 11 13:12 '#ib_redo17_tmp'
-rw-r----- 1 mysql mysql 3276800 Nov 11 13:12 '#ib_redo18_tmp'
-rw-r----- 1 mysql mysql 3276800 Nov 11 13:12 '#ib_redo19_tmp'
-rw-r----- 1 mysql mysql 3276800 Nov 11 13:12 '#ib_redo20_tmp'
-rw-r----- 1 mysql mysql 3276800 Nov 11 13:12 '#ib_redo21_tmp'
-rw-r----- 1 mysql mysql 3276800 Nov 11 13:12 '#ib_redo22_tmp'
-rw-r----- 1 mysql mysql 3276800 Nov 11 13:12 '#ib_redo23_tmp'
-rw-r----- 1 mysql mysql 3276800 Nov 11 13:12 '#ib_redo24_tmp'
-rw-r----- 1 mysql mysql 3276800 Nov 11 13:12 '#ib_redo25_tmp'
-rw-r----- 1 mysql mysql 3276800 Nov 11 13:12 '#ib_redo26_tmp'
-rw-r----- 1 mysql mysql 3276800 Nov 11 13:12 '#ib_redo27_tmp'
-rw-r----- 1 mysql mysql 3276800 Nov 11 13:12 '#ib_redo28_tmp'
-rw-r----- 1 mysql mysql 3276800 Nov 11 13:12 '#ib_redo29_tmp'
-rw-r----- 1 mysql mysql 3276800 Nov 11 13:12 '#ib_redo30_tmp'
-rw-r----- 1 mysql mysql 3276800 Nov 11 13:12 '#ib_redo31_tmp'
-rw-r----- 1 mysql mysql 3276800 Nov 11 13:12 '#ib_redo32_tmp'
-rw-r----- 1 mysql mysql 3276800 Nov 11 13:12 '#ib_redo33_tmp'
-rw-r----- 1 mysql mysql 3276800 Nov 11 13:12 '#ib_redo34_tmp'
-rw-r----- 1 mysql mysql 3276800 Nov 11 13:12 '#ib_redo35_tmp'
-rw-r----- 1 mysql mysql 3276800 Nov 11 13:12 '#ib_redo36_tmp'
-rw-r----- 1 mysql mysql 3276800 Nov 11 13:12 '#ib_redo37_tmp'
-rw-r----- 1 mysql mysql 3276800 Nov 11 13:14 '#ib_redo6'
-rw-r----- 1 mysql mysql 3276800 Nov 11 13:12 '#ib_redo7_tmp'
-rw-r----- 1 mysql mysql 3276800 Nov 11 13:12 '#ib_redo8_tmp'
-rw-r----- 1 mysql mysql 3276800 Nov 11 13:12 '#ib_redo9_tmp'
[root@jeames #innodb_redo]# pwd
/var/lib/mysql/#innodb_redo
```

附录图 1　MySQL8.0.31 重做日志文件

2. GIPK

从 MySQL 8.0.30 开始,MySQL 支持在 GIPK 模式下运行时生成的不可见主键。

在这种模式下运行时,对于任何没有显式主键创建的 InnoDB 表,MySQL 服务器会自动向表中添加生成的不可见主键(GIPK)。

新版本为我们提供了一个令人惊喜的特性 -(Generated Invisible Primary

Keys）简称 GIPK 。用一句概况：当开启 GIPK 模式后，MySQL 会在没有显示定义主键的 InnoDB 表上自动生成不可见的主键。如果没有主键，遇到 load data，大事务，ddl 等有大量表数据行扫描的行为时，会带来严重的主从延迟，给数据库稳定性和数据一致性带来隐患，那么 GIPK 解决了这个问题。

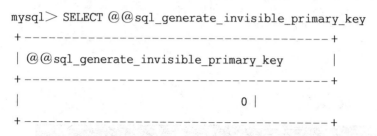

```
mysql> SELECT @@sql_generate_invisible_primary_key
+--------------------------------------+
| @@sql_generate_invisible_primary_key |
+--------------------------------------+
|                                    0 |
+--------------------------------------+
```

说明：GIPK 模式由 sql_generate_invisible_primary_key 服务器系统变量控制。默认情况下，该变量的值为 OFF，这意味着禁用了 GIPK 模式；要启用 GIPK 模式，请将变量设置为 ON。

接下来，演示下 GIPK 的特性，结果如附录图 2 所示：

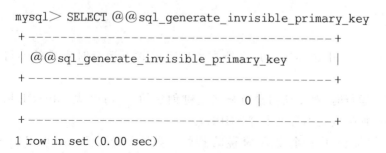

附录图 2　MySQL8.0.31 GIPK 的特性

2.1　GIPK 模式默认关闭

```
mysql> SELECT @@sql_generate_invisible_primary_key
+--------------------------------------+
| @@sql_generate_invisible_primary_key |
+--------------------------------------+
|                                    0 |
+--------------------------------------+
1 row in set (0.00 sec)
```

```
mysql> use jeames
mysql> CREATE TABLE auto_n1 (c1 VARCHAR(50), c2 INT)
```

##开启GIPK模式
```
mysql> SET sql_generate_invisible_primary_key = ON
mysql> SELECT @@sql_generate_invisible_primary_key
+--------------------------------------+
| @@sql_generate_invisible_primary_key |
+--------------------------------------+
|                                    1 |
+--------------------------------------+

mysql> CREATE TABLE auto_n2 (c1 VARCHAR(50), c2 INT);
```

2.2 使用 SHOW CREATE TABLE 查看表实际创建方式的差异

```
mysql> SHOW CREATE TABLE auto_n1\G
***************************1. row***************************
       Table: auto_n1
Create Table: CREATE TABLE 'auto_n1' (
  'c1' varchar(50) DEFAULT NULL,
  'c2' int DEFAULT NULL
) ENGINE = InnoDB DEFAULT CHARSET = utf8mb4 COLLATE = utf8mb4_0900_ai_ci
1 row in set (0.00 sec)

mysql> SHOW CREATE TABLE auto_n2\G
***************************1. row***************************
       Table: auto_n2
Create Table: CREATE TABLE 'auto_n2' (
  'my_row_id' bigint unsigned NOT NULL AUTO_INCREMENT /*! 80023 INVISIBLE */,
  'c1' varchar(50) DEFAULT NULL,
  'c2' int DEFAULT NULL,
  PRIMARY KEY ('my_row_id')
) ENGINE = InnoDB DEFAULT CHARSET = utf8mb4 COLLATE = utf8mb4_0900_ai_ci
1 row in set (0.00 sec)
```

说明：

（1）由于 auto_n2 没有用于创建它的 CREATE TABLE 语句指定的主键，因此 GIPK 模式使 MySQL 将不可见的列 my_row_id 和该列上的主键添加到此表中。

由于在创建 auto_n1 时禁用了 GIPK 模式，因此未在该表上执行此类添加。

（2）当服务器以 GIPK 模式将主键添加到表中时，列和键名称始终为 my_row_id。

因此，当启用 GIPK 模式时，除非表创建语句还指定了显式主键，否则不能创建具有名为 my_row_id 的列的表。

（在这种情况下，不需要为列或键命名 my_row_id。）

2.3 VISIBLE 和 INVISIBLE 之间切换

当 GIPK 模式生效时，生成的主键不能更改，只能在 VISIBLE 和 INVISIBLE 之间切换。要使 auto_n2 上生成的不可见主键可见，请执行以下 ALTER TABLE 语句：

mysql> ALTER TABLE auto_n2 ALTER COLUMN my_row_id SET VISIBLE

mysql> SELECT COLUMN_NAME, ORDINAL_POSITION, DATA_TYPE, COLUMN_KEY FROM INFORMATION_SCHEMA.COLUMNS
WHERE TABLE_NAME = "auto_n2"

要使生成的主键再次不可见，请执行

ALTER TABLE auto_1 ALTER COLUMN my_row_id SET invisible。

创建或导入使用 GIPK 模式的安装备份时，可以排除生成的不可见 PK 列和值。mysqldump 的 -skip 生成的不可见主键选项会导致 GIPK 信息被排除在程序的输出中。如果要导入包含 GIPK 键和值的转储文件，还可以使用 mysqlpump 中的 -skip 生成的不可见主键来抑制这些键和值（从而不导入）。

3. 多级别的 ORDER BY or LIMIT

在 MySQL 8.0.31 之前，带括号的查询表达式不允许多个级别的 ORDER BY or LIMIT 操作，查询会被拒绝。在 MySQL 8.0.31 及更高版本中，取消了此限制，并允许嵌套的带括号的查询表达式。支持的最大嵌套级别为 63；这是在解析器执行任何简化或合并之后。

以下是示例：

mysql> (SELECT 'a' UNION SELECT 'b' LIMIT 2) LIMIT 3
(
    (SELECT a, b, c FROM t ORDER BY a LIMIT 3) ORDER BY b LIMIT 2
) ORDER BY c LIMIT 1

4. innodb_doublewrite

系统变量支持 DETECT_ONLY 和 DETECT_AND_RECOVER 设置。使用该 DETECT_ONLY 设置,数据库页面内容不会写入双写缓冲区,并且恢复不会使用双写缓冲区来修复不完整的页面写入。此轻量级设置仅用于检测不完整的页面写入。该 DETECT_AND_RECOVER 设置等同于现有 ON 设置。

5. Mysqldump

Mysqldump 执行全表扫描,这意味着它的查询通常会超过 long_query_time 对常规查询有用的设置。从 MySQL 8.0.30 开始,如果要从慢查询日志中排除大部分或全部 Mysqldump 产生的慢查询,可以设置 Mysqldump 的 --mysqld-long-query-time 命令行选项,将系统变量的 session 值更改为更高的值。

附录 2　性能优化

1. 统计信息

MySQL 执行 SQL 会经过 SQL 解析和查询优化的过程，解析器将 SQL 分解成数据结构并传递到后续步骤，查询优化器发现执行 SQL 查询的最佳方案、生成执行计划。查询优化器决定 SQL 如何执行，依赖于数据库的统计信息。

MySQL 统计信息的存储分为两种，非持久化和持久化统计信息。

（1）非持久化

非持久化统计信息存储在内存里，如果数据库重启，统计信息将丢失。

mysql> SELECT @@INNODB_STATS_PERSISTENT

非持久化统计信息的缺点显而易见，数据库重启后如果大量表开始更新统计信息，会对实例造成很大影响，所以目前都会使用持久化统计信息。

（2）持久化

5.6.6 开始，MySQL 默认使用了持久化统计信息，即

INNODB_STATS_PERSISTENT = ON

持久化统计信息保存在表 mysql.innodb_table_stats 和 mysql.innodb_index_stats

mysql>SELECT @@INNODB_STATS_AUTO_RECALC

2. 统计信息不准确的处理

我们查看执行计划，发现未使用正确的索引，如果是 innodb_index_stats 中统计信息差别较大引起，可通过以下方式处理：

（1）手动更新统计信息

执行过程中会加读锁：

ANALYZE TABLE TABLE_NAME;

（2）增加数据页

如果更新后统计信息仍不准确，可考虑增加表采样的数据页。

两种方式可以修改：

①全局变量 INNODB_STATS_PERSISTENT_SAMPLE_PAGES,默认为 20;

②单个表可以指定该表的采样:ALTER TABLE TABLE_NAME STATS_SAMPLE_PAGES=40;

经测试,此处 STATS_SAMPLE_PAGES 的最大值是 65 535,超出会报错。

某些情况下(如数据分布不均)仅仅更新统计信息不一定能得到准确的执行计划,只能通过 index hint 的方式指定索引。

新版本 8.0 增加了直方图功能,MYSQL 是从 8.0 版本开始的(官方版本 8.0.3),引入的列的直方图的信息,从而来帮助优化器来选择相对更好的执行计划。

mysql 的直方图目前有 2 种形式:singleton、equi-height。

Singleton:特点是每个值一个 bucket,每个 bucket 保存的信息为列的值和该值出现的频率。使用场景为等值查询和范围查询。

Equi-height:特点是每个桶里面会保存多个列的值,每个 bucket 保存的信息为:桶中列的最大值和最小值,累计出现该值的频率,NDV (number of distinct value)等信息,使用场景为范围查询。

直方图收集的方法:

mysql> analyze table t_trx_attendance UPDATE HISTOGRAM ON attend_start_time WITH 8 BUCKETS

3. 执行计划查看

我们经常会使用 Explain 去查看执行计划,下面我们就详细讨论下 Explain 中的"Type"和"Extra"。

3.1 Type

Explain 中的"Type",MySQL 的官网解释为:连接类型(the join type)。它描述了找到所需数据使用的扫描方式。

最为常见的扫描方式有:
system:系统表,少量数据,往往不需要进行磁盘 IO
const:常量连接
eq_ref:主键索引(primary key)或者非空唯一索引(unique not null)等值扫描
ref:非主键非唯一索引等值扫描
range:范围扫描
index:索引树扫描
ALL:全表扫描(full table scan);
上面各类扫描方式由快到慢:

system > const > eq_ref > ref > range > index > ALL

3.2 Extra

Extra 的值有 NULL、Using index、Using where、Using index condition、Using filesort、Using temporary。

3.2.1 Using where

mysql> explain select * from account_user_base where id > 4

Extra 为 Using where 说明,SQL 使用了 where 条件过滤数据。

3.2.2 Using index

mysql> explain select id from account_user_base

Extra 为 Using index 说明,SQL 所需要返回的所有列数据均在一棵索引树上,而无需访问实际的行记录。

3.2.3 Using index condition

mysql> explain select * from account_user_security t1 account_user_base t2 where t1.user_id = t2.id

Extra 为 Using index condition 说明,确实命中了索引,但不是所有的列数据都在索引树上,还需要访问实际的行记录。

3.2.4 Using filesort

mysql> explain select id from account_user_base order by nick_name

Extra 为 Using filesort 说明,得到所需结果集,需要对所有记录进行文件排序。

典型的,在一个没有建立索引的列上进行了 order by,就会触发 filesort,常见的优化方案是,在 order by 的列上添加索引,避免每次查询都全量排序。

3.2.5 Using temporary

mysql> explain select nick_name, COUNT(*) from account_user_base GROUP BY nick_name order by nick_name

Extra 为 Using temporary 说明,需要建立临时表(temporary table)来暂存中间结果。这类 SQL 语句性能较低,往往也需要进行优化。典型的,group by 和 order by 同时存在,且作用于不同的字段时,就会建立临时表,以便计算出最终的结果集。

附录 3　MySQL 技能树

1. 关于 MySQL 服务器,以下说法错误的是:(C)
   A. MySQL 可以设置监听端口、地址和最大连接数
   B. MySQL 的每个数据表可以指定存储引擎
   C. MySQL 的服务器和客户端必须运行在不同的服务器上
   D. MySQL 的超级用户默认为 root
2. 下列数据库产品中,哪一个不是关系型数据库?(B)
   A. PostgreSQL
   B. redis
   C. MySQL
   D. H2
3. Jeames 需要登录服务器上的 goods 数据库,这台服务器地址是 10.70.1.11,MySQL 端口是 3306。用户名是 jem,口令是 it。那么 Jeames 需要如何操作?(A)
   A. 在终端输入
   mysql -h 10.70.1.11 -p -ujem goods
   出现口令输入提示时输入口令。
   B. 在终端输入
   mysql tcp://jem@10.70.1.11:3306/goods
   出现口令输入提示时输入口令。
   C. 在终端输入
   mysql -h 10.70.1.11 -p3306 -ujem goods
   出现口令输入提示时输入口令。
   D. 在终端输入
   mysql -ujem 10.70.1.11/goods
   出现口令输入提示时输入口令。
4. jeames 想要在 goods 数据库创建一个 category 表,管理商品的类别,那么正确的建表语句应该是:(C)
   A.
   DROP TABLE category
   (
       id   INT PRIMARY KEY

```
        name    VARCHAR(30)
        remark  VARCHAR(100)
)
```
B.
```
MAKE TABLE category
(
        id      INT PRIMARY KEY
        name    VARCHAR(30)
        remark  VARCHAR(100)
)
```
C.
```
CREATE TABLE category
(
        id      INT PRIMARY KEY
        name    VARCHAR(30)
        remark  VARCHAR(100)
)
```
D.
```
ADD TABLE category
(
        id      INT PRIMARY KEY
        name    VARCHAR(30)
        remark  VARCHAR(100)
);
```

5. jeames 想要删除数据库中的 category 表，正确的操作是？（C）

    A. truncate table category

    B. delete table where name = 'category'

    C. drop table category

    D. remove table category

6. 关于 MySQL 的存储引擎，下列说法错误的是：(D)

    A. InnoDB 支持事务，有更好的并发能力

    B. MyISAM 不支持事务和外键，结构简单，可以压缩为只读状态

    C. Memory 引擎将数据保存在内存中，重启会丢失数据，读速度快很快，适合作为会话表和缓存表

    D. 临时表默认使用 Memory 引擎

7. 订单表建表语句如下：

create table orders (
    id int primary key auto_increment
    item_id int
    amount int
    unit_price decimal(12,4)
    price decimal(12,4)
    description varchar(2000)
    ts timestamp default now()
)

jeames 需要添加一个字段，用来保存订单的相关图片，由于特殊的业务需要，这些图片必须保存在数据库中，图片的大小不超过 100 K。

那么应该使用以下哪个建表语句：(A)

    A.

create table orders (
    id int primary key auto_increment
    item_id int
    amount int
    unit_price decimal(12,4)
    price decimal(12,4)
    description varchar(2000)
    picture blob
    ts timestamp default now()
)

    B.

create table orders (
    id int primary key auto_increment,
    item_id int
    amount int
    unit_price decimal(12,4)
    price decimal(12,4)
    description varchar(2000)
    picture text
    ts timestamp default now()
)

C.
```
create table orders (
    id int primary key auto_increment
    item_id int
    amount int
    unit_price decimal(12,4)
    price decimal(12,4)
    description varchar(2000)
    picture binary(100000)
    ts timestamp default now()
)
```
D.
```
create table orders (
    id int primary key auto_increment
    item_id int
    amount int
    unit_price decimal(12,4)
    price decimal(12,4)
    description varchar(2000)
    picture varbinary(100000)
    ts timestamp default now()
)
```

在 mysql 中，存储图片用 BLOB 类型。

BLOB 类型是一种特殊的二进制类型，可以存储数据量很大的二进制数据，包括图片、视频 MySQL 的四种 BLOB 类型，大小（单位：字节）。

TinyBlob 最大 255

Blob 最大 65 K

MediumBlob 最大 16 M

LongBlob 最大 4 G

BINARY 和 VARBINARY 类型类似于 CHAR 和 VARCHAR，不同的是它们包含二进制字节字符串。

8. jeames 希望从 orders 表

```
create table orders
(
    id            int primary key auto_increment
```

```
    item_id        int
    amount         int
    unit_price     decimal(12,4)
    price          decimal(12,4)
    description    varchar(2000)
    ts             timestamp default now()
    deal           bool       default false
)
```

查询 2022 年 10 月 25 日下单的所有单价低于 30 的订单 id,查询正确的是?(A)

A.
```
select id
from orders
where date(ts) = '2022-10-25'
  and unit_prise < 30
```

B.
```
select id
from orders
which date(ts) = '2022-10-25'
   or unit_prise < 30
```

C.
```
select id
from (select * from orders where date(ts) = '2022-10-25') as o
where unit_prise < 30
```

D.
```
select id
from orders
if date(ts) = '2022-10-25' or unit_prise < 30
```

9. 关于 mysqlshow 工具,以下说法正确的是:(C)

(1)mysqlshow 命令主要用来查看 MySQL 中存在的数据库和数据表,以及数据表中的字段和索引等信息

(2)可以在 /etc/my.cnf /etc/mysql/my.cnf /usr/local/etc/my.cnf ~/.my.cnf 中的某一个配置文件中配置连接选项

MySQLShow 会按顺序加载,重复项则后面的覆盖前面的。

(3)--count 选项可以查看各数据库中表的数量和数据行数

(4)传入数据库名可以查看指定数据库的各表行数和列数

（5）传入库名和表名可以查看指定数据表的列数和行数

（6）指定 -k 选项可以查看索引的信息

  A. 2，3，4，5

  B. 2，3，5，6

  C. 全部都对

  D. 全都不对

  mysqlshow 命令主要用来查看 MySQL 中存在的数据库和数据表，以及数据表中的字段和索引等信息，使用格式如下：

shell> mysqlshow [options] [db_name [tbl_name [col_name]]]

  其中，options 的取值可以输入如下命令进行查阅。

mysqlshow --help

  可以参考 MySQL 官方文档。

网址为 https://dev.mysql.com/doc/refman/8.0/en/mysqlshow.html。

（1）查看 MySQL 中所有的数据库。

mysqlshow -uroot -p

（2）查看每个数据库中数据表的数量和数据的总条数信息

mysqlshow -uroot -p --count

（3）查看 goods 数据库下每个表中字段的数量和表中记录的数量

mysqlshow -uroot -p goods --count

（4）查看 goods 数据库下 t_goods 数据表的信息

mysqlshow -uroot -p goods t_goods --count

（5）查看 goods 数据库下 t_goods 数据表中的所有索引

mysqlshow -uroot -p goods t_goods -k

（6）显示 t_goods 数据表的状态信息

mysqlshow -uroot -p goods t_goods -i

10. 关于 MySQL 的自增字段，错误的说法是：(D)

  A. 自增字段必须是主键

  B. 插入操作失败，自增计数仍然会被递增，下次操作使用下一个整数

  C. 自增字段默认从 1 开始

  D. 自增字段必须名为 id

11. jeames 要查询 goods 表 sql

create table goods(

id int primary key auto_increment

category_id int

category varchar(64)

```
name varchar(256)
price decimal(12,4)
stock int
upper_time timestamp
```
)中价格在 2000 到 3000 之间(包含 2000 和 3000)的数据,以下查询错误的是:(A)

　　A. SELECT * FROM goods HAVING price BETWEEN 2000 AND 3000
　　B. SELECT * FROM goods WHERE price BETWEEN 2000 AND 3000
　　C. SELECT * FROM goods WHERE price >= 2000 AND price <= 3000
　　D. SELECT * FROM goods WHERE not (price < 2000 or price > 3000)

12. jeames 要从 employee 表 sql
```
create table employee
(
id      serial primary key
name    varchar(256)
dept    varchar(256)
salary  decimal(12,4)
)
```
中得到每月工资开支超过五万的部门,这个查询应该怎么写?(C)

　　A. select dept from employee where sum(salary) > 50000 order by dept
　　B. select dept from employee group by dept where sum(salary) > 50000
　　C. select dept from employee group by dept having sum(salary) > 50000
　　D. select dept from employee where sum(salary) > 50000 group by dept

13. 现有员工信息表如下:
```
create table employee
(
    id      serial primary key
    name    varchar(256)
    dept    varchar(256)
    salary  money
)
```
下面哪条查询,可以给出每个部门工资最高的员工的 id,name,dept,salary 四项信息?(D)

　　A.
```
select id, name, dept, max(salary)
from employee
```

group by dept

B.

select id, name, dept, max(salary)

from employee

where salary = max(salary)

group by dept

C.

select id, name, dept, max(salary)

from employee

group by dept, id, name

D.

select l.id, l.name, l.dept, l.salary

from employee as l

  join (select max(salary) as salary, dept

   from employee

   group by dept) as r

   on l.dept = r.dept and l.salary = r.salary

14. jeames 想要给 goods 数据库加上慢查询日志,将其保存到 /data/log/mysql/goods/。他需要执行下列哪些操作?

(1)执行 shell 命令 mkdir -p /data/log/mysql/goods/

(2)编辑 my.cnf 的 [mysqld]节,设置 slow_log = 1

(3)slow_log_file = /data/log/mysql/goods/slow_statement.log

(4)log_output = FILE

(5)重启 MySQL 服务

(6)备份数据库

(7)恢复数据库

请在以下选项中选择（A）

 A. 1,2,3

 B. 1,2,3,4,5

 C. 1,2,4,5

 D. 1,2

15. 管理员要给用户 jeames 授权,允许他查询 emplyee 表,应用哪一条语句?（B）

 A. grant query on table employee to jeames

 B. grant select on table employee to jeames

 C. grant read on table employee to jeames

D. grant all on table employee to jeames

16. Shop 表的部分字段如下：

create table shop (

    id int primary key auto_increment

    description varchar(6000)

)

现在 jeames 要给 description 字段加上全文索引，正确的语句是？（D）

    A. alter table shop create fulltext(description)

    B. alter table shop add index fulltext(description)

    C. alter table shop alter description add fulltext(description)

    D. alter table shop add fulltext(description)

17. 交易过程 trade 中有一个遍历交易项的循环，当累计的总交易额 @total_price 超过 30 000，这个循环就结束

    -- ...

trade：LOOP

    -- ...

    if @total_price > 30 000 then

        LEAVE trade

    else

        ITERATE trade

    end if

    -- ...

end LOOP trace

现在，jeames 想要用 REPEATE 简化这个 LOOP 循环，他应该怎么做？（A）

    A.

trade：REPEATE

    --省略交易过程

    UNTIL @total_price > 30 000

END REPATE trade

    B.

REPEATE @total_price > 30 000

    --省略交易过程

END REPATE

    C.

trade：REPEATE

```
        IF @total_price > 30 000
            --省略交易过程
    END REPATE trade
D.
    trade：REPEATE
        UNTIL @total_price > 30 000
            --省略交易过程
    END REPATE trade
```

在 REPEAT 语句中不管是否满足给定条件，首先会执行一次 statements，然后再在 UTILE 中判断给定的条件是否成立，如果条件不成立会继续执行，如果条件成立则退出 REPEAT 循环。

18. 在主从复制一节里，jeames 实现了一个点对点的主从复制架构，其中 standby 是 trade 的从库，现在，jeames 要添加一个名为 backup 的新的复制节点，这个节点的同步进度要比 trade 晚半小时。这时应该怎么做？（C）

    A. 所有都不对

    B. 按添加从库 一节的步骤添加 backup 节点，在主库重启推送之前，执行 SET BACKUP_DELAY=1 800

    C. 按添加从库 一节的步骤添加 backup 节点，在设置 backup 的主库订阅信息时，加入 MASTER_DELAY=1 800

    D. 按添加从库 一节的步骤添加 backup 节点，在设置 backup 的主库订阅信息时，加入 MASTER_DELAY=0.5

19. Goods 数据库近期在每日高峰时段很慢，jeames 想初步查看一下数据库的工作状态，他应该执行：（A）

    A. show global status

    B. show status

    C. show local status

    D. show session status

--查看 MySQL 本次启动后的运行时间（单位：秒）
show status like 'uptime'

--查看 select 语句的执行数
show [global] status like 'com_select'

--查看 insert 语句的执行数
show [global] status like 'com_insert'

--查看 update 语句的执行数

show [global] status like 'com_update'

--查看 delete 语句的执行数

show [global] status like 'com_delete'

--查看试图连接到 MySQL(不管是否连接成功)的连接数

show status like 'connections'

--查看线程缓存内的线程的数量

show status like 'threads_cached'

--查看当前打开的连接的数量

show status like 'threads_connected'

--查看当前打开的连接的数量

show status like 'threads_connected'

--查看创建用来处理连接的线程数。如果 Threads_created 较大,可能要增加 thread_cache_size 值

show status like 'threads_created'

--查看激活的(非睡眠状态)线程数

show status like 'threads_running'

--查看立即获得的表的锁的次数

show status like 'table_locks_immediate'

--查看不能立即获得的表的锁的次数。如果该值较高,并且有性能问题,应首先优化查询,然后拆分表或使用复制

show status like 'table_locks_waited'

--查看创建时间超过 slow_launch_time 秒的线程数

show status like 'slow_launch_threads'

--查看查询时间超过 long_query_time 秒的查询的个数

show status like 'slow_queries'

-- QPS(每秒 Query 量)

-- QPS = Questions(or Queries) / seconds

show global status like 'Question%'

-- TPS(每秒事务量)

-- TPS = (Com_commit + Com_rollback) / seconds

show global status like 'Com_commit'

show global status like 'Com_rollback'

-- key Buffer 命中率

show global status like 'key%'

key_buffer_read_hits = (1 - key_reads / key_read_requests) * 100%
key_buffer_write_hits = (1 - key_writes / key_write_requests) * 100%

-- InnoDB Buffer 命中率

show status like 'innodb_buffer_pool_read%'

innodb_buffer_read_hits = (1 - innodb_buffer_pool_reads / innodb_buffer_pool_read_requests) * 100%

-- Query Cache 命中率

show status like 'Qcache%'

Query_cache_hits = (Qcahce_hits / (Qcache_hits + Qcache_inserts)) * 100%

-- Table Cache 状态量

show global status like 'open%'

比较 open_tables 与 opend_tables 值

-- Thread Cache 命中率

show global status like 'Thread%'
show global status like 'Connections'

Thread_cache_hits = (1 - Threads_created / connections) * 100%

--锁定状态

mysql> show global status like '%lock%'

Table_locks_waited/Table_locks_immediate = 0.3% 如果这个比值比较大,说明表锁造成的阻塞比较严重

Innodb_row_lock_waits innodb 行锁,很可能是间隙锁造成的

-- Tmp Table 状况(临时表状况)

show status like 'Create_tmp%'

Created_tmp_disk_tables/Created_tmp_tables 比值最好不要超过 10%,如果 Created_tmp_tables 值比较大

可能是排序句子过多或者是连接句子不够优化。

-- Binlog Cache 使用状况

show status like 'Binlog_cache%'

如果 Binlog_cache_disk_use 值不为 0，可能需要调大 binlog_cache_size 大小

-- Innodb_log_waits 量

show status like 'innodb_log_waits'

Innodb_log_waits 值不等于 0，表明 innodb log buffer 因为空间不足而等待

20. jeames 需要删除数据分析库中 orders 表的数据，orders 按时间分区，因为是分析部门离线使用，不需要考虑并发，下列哪些操作可以更快删除这些数据？（A）

(1) 使用 truncate from orders

(2) 可以执行 ALTER TABLE orders DROP PARTITION partition_name; 删除指定分区

(3) 去掉唯一约束然后 delete from orders where 1=1

(4) 使用可写游标，一次一万行滚动删除

(5) drop table orders 删除后重建

  A. 1，2，5

  B. 2，3，4

  C. 2，3

  D. 1，2，3，4